Ausgeschieden im Jahr 2025

THERMODYNAMICS
OF MATERIALS

THERMODYNAMICS OF MATERIALS
A Classical and Statistical Synthesis

JOHN B. HUDSON

Rensselaer Polytechnic Institute

A Wiley-Interscience Publication
JOHN WILEY & SONS, INC.
New York / Chichester / Brisbane / Toronto / Singapore

The text is printed on acid-free paper.

Copyright ©1996 by John Wiley & Sons, Inc.

All rights reserved. Published simultaneously in Canada.

Reproduction or translation of any part of this work beyond that permitted by Section 107 or 108 of the 1976 United States Copyright Act without the permission of the copyright owner is unlawful. Requests for permission or further information should be addressed to the Permissions Department, John Wiley & Sons, Inc., 605 Third Avenue, New York, NY 10158-0012.

Library of Congress Cataloging in Publication Data:

Hudson, John B.
    Thermodynamics of materials : a classical and statistical synthesis/by John Hudson.
      p.   cm.
    Includes index.
    ISBN 0-471-31143-X (alk. paper)
    1. Materials--Thermal properties. 2. Thermodynamics. I. Title.
TA418.52.H83   1996
620.1'1296--dc20
                                                  95-20224

Printed in the United States of America

10 9 8 7 6 5 4 3 2 1

To my wife and children
Carol, Rob, Dave, and Jean

CONTENTS

PREFACE xv

PART I REVIEW OF CLASSICAL THERMODYNAMICS

1 INTRODUCTION   3

   Definitions / 4
   Bibliography / 8

2 THE LAWS OF CLASSICAL THERMODYNAMICS   9

   The Zeroth Law / 9
   Work and the First Law / 10
   Heat Flow and the Second Law / 12
   The Absolute Temperature / 13
   The Third Law / 15
   Bibliography / 16
   Problems / 16

3 CRITERIA FOR EQUILIBRIUM   19

   The Entropy Principle / 19
   Definitions / 20
   The Basic Equilibrium Postulate / 21
   Additional State Functions / 22
   Criteria for Equilibrium / 23
   The Isolated System / 24

The Closed Isothermal System / 25
The Closed Isobaric System / 26
The Open Isothermal System / 26
Bibliography / 27
Problems / 28

4 USEFUL MATHEMATICAL RELATIONS 29

Partial Derivatives / 29
Relation to Measurable Properties / 30
Evaluation of Partial Derivatives / 31
The Cross-Differentiation Identity / 32
Maxwell's Relations / 34
Integration of the State Function Equations / 35
Bibliography / 35
Problems / 36

5 GENERAL THEORY OF PHASE STABILITY 39

General Relations / 39
Degrees of Freedom / 40
The Gibbs–Duhem Equation / 40
Equilibrium in Multiphase Systems / 41
Thermal Equilibrium / 42
Hydrostatic Equilibrium / 42
Distributive Equilibrium / 43
Chemical Equilibrium / 44
General Criteria for Equilibrium / 46
The Gibbs Phase Rule / 47
Bibliography / 48
Problems / 48

PART II FUNDAMENTALS OF STATISTICAL THERMODYNAMICS

6 BASIS OF STATISTICAL THERMODYNAMICS 53

What Is Statistical Thermodynamics? / 53
Basic Approach / 54
Relation of Macroscopic to Microscopic Descriptions / 56
Ensembles / 56
Types of Ensembles / 59
Postulates / 60
Bibliography / 63
Problems / 63

7 EVALUATION OF PROBABILITIES 65

Application of Postulates / 65
The Microcanonical Ensemble / 66
The Canonical Ensemble / 66
The Grand Canonical Ensemble / 70
Bibliography / 72
Problems / 73

8 STATISTICAL MECHANICAL CRITERIA FOR EQUILIBRIUM 75

The Isolated System and the Function S' / 75
The Closed Isothermal System and the Function F' / 78
The Open Isothermal System and the Function $(PV)'$ / 79
Bibliography / 81
Problems / 81

9 THE CONNECTION BETWEEN STATISTICAL THERMODYNAMICS AND CLASSICAL THERMODYNAMICS 83

The Canonical Ensemble and β / 84
Evaluation of the State Functions in Terms of Q / 87
The Grand Canonical Ensemble and γ / 90
Evaluations of the State Functions in Terms of Ξ / 93
Fluctuations / 95
Bibliography / 98
Problems / 98

10 EVALUATION OF THE ALLOWED ENERGIES 101

Models / 101
Quantum Mechanics / 102
Translational Motion / 103
Rotational Motion / 104
Motion in a Potential Field / 105
Degeneracy and Interparticle Effects / 106
Independent Particle Systems / 108
Relation of System Energy to Particle Energies / 108
Distinguishable Particle Systems / 110
Indistinguishable Particle Systems / 111
Independence of Modes of Energy Storage / 112
Bibliography / 113
Problems / 114

PART III SINGLE-COMPONENT SYSTEMS

11 CLASSICAL THERMODYNAMICS OF ONE-COMPONENT SYSTEMS   119

Free Energy Surfaces / 119
Temperature Dependence of the Thermodynamic Functions / 122
Pressure Dependence of the Thermodynamic Functions / 126
The One-Component Phase Diagram / 127
Molar Properties / 129
The Clapeyron Equation / 130
Evaluation of the State Functions / 133
Bibliography / 133
Problems / 133

12 THE MONATOMIC IDEAL GAS   135

The Model / 135
Number of Available States / 136
Evaluation of q / 138
Evaluation of the Partition Function / 141
Evaluation of the Thermodynamic Functions / 142
Electronic Excitation / 144
The Zero of Energy / 145
Complete Expressions for the Thermodynamic Functions / 146
Bibliography / 146
Problems / 147

13 THE POLYATOMIC IDEAL GAS   149

The Model / 149
Evaluation of q / 152
Evaluation of the Thermodynamic Functions / 156
Polyatomic Molecules / 158
The Grand Canonical Ensemble / 158
Bibliography / 160
Problems / 160

14 THE EINSTEIN MODEL OF THE SOLID   163

The Einstein Model / 163
Evaluation of the Partition Function / 166
Limiting Values of q_v / 168

CONTENTS **xi**

Evaluation of the Thermodynamic Functions / 169
High and Low Temperature Limits / 171
Bibliography / 173
Problems / 173

15 THE DEBYE MODEL OF THE SOLID 175

The Debye Model / 175
Evaluation of the Partition Function / 178
Evaluation of the Thermodynamic Functions / 181
Relation of Θ_D to Crystal Properties / 183
High and Low Temperature Limits / 184
Bibliography / 186
Problems / 186

16 SIMPLE LIQUIDS 187

The Model / 187
Evaluation of the Partition Function / 190
Evaluation of the Thermodynamic Functions / 191
Evaluation of Parameters / 191
Bibliography / 194
Problems / 195

17 STATISTICAL THERMODYNAMICS OF PHASE EQUILIBRIUM IN ONE-COMPONENT SYSTEMS 197

Solid–Vapor Equilibrium / 197
Liquid–Vapor Equilibrium / 198
The Triple Point / 200
Solid–Liquid Equilibrium / 200
A Numerical Example / 201
Bibliography / 206
Problems / 206

PART IV MULTICOMPONENT SYSTEMS

18 CLASSICAL THERMODYNAMICS OF MULTICOMPONENT SYSTEMS 211

Activity / 211
Molar Properties / 212
Partial Molar Properties / 213

Relation of Partial to Total Molar Properties / 216
Calculation of Partial Properties from Total Properties / 217
Summary / 219
Bibliography / 220
Problems / 220

19 CLASSICAL THERMODYNAMICS OF SOLUTIONS   223

Formation of a Solution / 223
Ideal Gas Mixtures / 224
Multicomponent Condensed Phases / 227
The Ideal Solution / 228
Dilute Solutions / 229
Concentrated Solutions / 232
Excess Functions / 232
Bibliography / 234
Problems / 234

20 LATTICE STATISTICS   235

The Ideal Lattice Gas / 235
The Einstein Crystal with Vacancies / 236
Evaluation of the Thermodynamic Functions / 237
The Langmuir Model of Adsorption / 240
The Two-Dimensional Pressure / 243
Evaluation of the Thermodynamic Functions / 244
The Langmuir Adsorption Isotherm / 245
Bibliography / 247
Problems / 248

21 THE LATTICE GAS WITH INTERACTIONS   249

The Model / 249
Evaluatio n of the Partition Function / 251
The Bragg–Williams Approximation / 252
The Quasichemical Model / 256
Bibliography / 258
Problems / 258

22 STATISTICAL THERMODYNAMIC TREATMENT
OF SOLUTIONS   261

The Model / 261
Solid Solutions / 263
The Ideal Solid Solution / 264
The Bragg–Williams Model and Regular Solutions / 267

The Quasichemical Model / 272
Liquid Solutions / 277
The Partition Function / 278
Ideal Liquid Solutions / 279
Regular Liquid Solutions / 280
Bibliography / 281
Problems / 282

23 PHASE EQUILIBRIUM IN MULTICOMPONENT SYSTEMS 283

The Model / 283
The Two-Component Ideal System / 284
A Numerical Example / 288
Gas-Phase–Condensed-Phase Equilibrium / 290
Two-Component Nonideal Systems / 291
Phase Separation / 291
A Numerical Example / 292
Solid–Liquid Equilibrium in Nonideal Systems / 293
A Numerical Example / 296
Bibliography / 300
Problems / 300

24 CHEMICAL EQUILIBRIUM 303

The Equilibrium Constant / 303
Dissociation of a Diatomic Molecule / 304
Isotopic Equilibrium / 307
General Gas-Phase Reactions / 309
Heterophase Reactions / 310
Bibliography / 313
Problems / 313

PART V QUANTUM SYSTEMS

25 THE PERFECT ELECTRON GAS 317

The Model / 317
Number of Available States / 318
Evaluation of the Partition Function / 320
Average Number of Particles Per State / 321
Evaluation of the Thermodynamic Functions at $0\,\mathrm{K}$ / 322
The Fermi Energy and the Work Function / 326
Temperature Dependence of the Thermodynamic Functions / 327
Bibliography / 330
Problems / 330

xiv CONTENTS

26 BLACKBODY RADIATION   **331**

The Model / 331
Evaluatio n of the Partition Function / 332
Evaluation of the Thermodynamic Functions / 335
The Spectral Energy Distribution / 337
Bibliography / 340
Problems / 340

APPENDIX A: FUNDAMENTAL CONSTANTS AND CONVERSIONS   **341**

APPENDIX B: OTHER ENSEMBLES   **342**

APPENDIX C: THERMODYNAMIC DATA FOR ONE-COMPONENT SYSTEMS   **344**

APPENDIX D: VAPOR PRESSURE RELATIONS   **346**

APPENDIX E: MODEL POTENTIALS   **348**

APPENDIX F: SPECTROSCOPIC DATA FOR DIATOMIC AND POLYATOMIC MOLECULES   **350**

APPENDIX G: HYPERBOLIC FUNCTIONS   **352**

APPENDIX H: THE DEBYE FUNCTION AND THE DEBYE TEMPERATURE   **354**

APPENDIX I: THERMODYNAMIC DATA FORCHEMICAL REACTIONS   **356**

APPENDIX J: WORK FUNCTIONS AND FERMI ENERGIES   **358**

INDEX   **359**

PREFACE

Many years ago, when I first started teaching the course which has evolved into the material presented in this text, I was sitting at home, reading the text that I was then using in preparation for my next lecture. My young son, David, toddled into the room, looked up and asked "What ya readin', dad?" Looking down at him, I intoned "Statistical Thermodynamics." His eyes grew wide, and he asked "Does that make brooms sweep all by themselves?" This response, while it implies a greater familiarity with Dukas' "The Sorcerer's Apprentice" than with statistical thermodynamics, is typical of the response of most students on their initial exposure to the subject. It is viewed as something mysterious, perhaps magical, with no obvious connection to the world of reality. In this text, I have tried to dispel this negative image and to show the natural connection between the classical and statistical approaches to thermodynamics, both as a means of obtaining useful information about real systems and as a way of showing the relation between the molecular level properties of systems and their properties on a macroscopic scale. The basic aim throughout has been to introduce the rigorous, general relations that arise from classical thermodynamic considerations, and which are system-independent, and then to use statistical thermodyamic relations in connection with molecular-level models of the specific systems involved to calculate the expected values of the macroscopic thermodynamic parameters of the systems.

The general organization of the text reflects this underlying aim. We begin with a review of classical thermodynamics, necessarily brief, because there is much material to be covered. I have assumed a working knowledge of classical thermodynamics, based on previous exposure of the student at the undergraduate level. This section is followed by an introduction to statistical thermodynamics which assumes no previous knowledge of the subject. In this development, the subject is introduced using the ensemble concept of Gibbs. The development of the partition functions for the various ensembles follows a method that is more intuitive than mathematical, because it is all too easy to lose sight of the physics of the situation in the more complex mathematics of, say, the method of undetermined multipliers. The remainder of this section develops the relations connecting the classical and statistical approaches and introduces the simplifications that will be used in treating practical systems.

The next two sections of the text follow the process of first developing general, classical relations for various classes of systems, then, through appropriate models, developing expressions in terms of molecular-level parameters that can be used to evaluate these general relations. Finally, these relations are used to characterize the behavior of systems with respect to phase and chemical equilibrium. The text closes with a short section on systems in which we must take explicit account of quantum mechanical restrictions on the number of particles per microstate.

In all cases the models chosen represent the simplest model in terms of mathematical complexity that will demonstrate the general features of the observed behavior of the system under study. We will deal with harmonic oscillators, rigid rotors, and interatomic potentials that can be described by the simple Lennard-Jones 6-12 model. Again, this choice has been made in order that the physical relationships between molecular-level and macroscopic-level behavior will not be obscured by complex mathematics.

I have tried throughout to stick to basic systems of general interest because of the limitations imposed by the amount of class time available in the semester. As a result, many interesting phenomena that are well treated by statistical thermodynamics, such as polymer elasticity, surface thermodynamics, electrolyte solutions, and order–disorder phenomena in solids, have not been treated.

For several years I have been using the draft version of this work as the primary text in the first-year graduate course "Advanced Thermodynamics" that I teach annually in the Materials Science and Engineering Department at Rensselaer. I generally cover one chapter of this work per 90-minute class period, in a format that includes an initial summary of the material, in which I assume that the students have read the appropriate chapter in the text prior to class time. This is followed by a question period in which points made in the summary can be clarified or expanded upon, and I finish with an interactive session in which students work in small groups to do one of the problems at the end of the chapter. My experience is that, in the 14-week semester we currently have, I cannot get quite through all of the material in the text. In particular, Chapters 25 and 26 are often slighted. It may be possible to shorten the time spent on Parts I or II, depending on the background of the class, or to omit topics that may be of marginal interest in some disciplines such as that on diatomic or polyatomic gases or on adsorbed gases.

In developing this text, I have, of course, made use of the work of many previous authors. The bibliographic citations at the end of each chapter reflect the sources that I have used. I owe a particular debt to the classic *Introduction to Statistical Thermodynamics* by Terrell Hill, and have also relied heavily on the work of the same title by Eldon Knuth. I have, in the past, used both of these texts in my own teaching, but I have abandoned them primarily because of my desire to include more material on condensed phase (especially multicomponent condensed phase) systems and on the application of thermodynamics to phase and chemical equilibrium, rather than because of any inherent deficiency on the part of these works.

I would like to thank Professor John W. Halloran, of the Department of Materials Science and Engineering at the University of Michigan for a critical review of the manuscript and for many helpful comments.

Troy, New York   JOHN B. HUDSON

PART I

REVIEW OF CLASSICAL THERMODYNAMICS

CHAPTER 1

INTRODUCTION

The basic purpose of this course is to develop the techniques of classical and statistical thermodynamics in tandem and to show how these techniques can be used to describe the equilibrium properties and behavior of systems of interest to materials scientists and engineers.

We will introduce the concepts involved in classical and statistical thermodynamics, or statistical mechanics, by comparing and contrasting the *macroscopic* approach to the description of systems, used in classical thermodynamics, with the *microscopic* approach used in statistical thermodynamics. We will see that the behavior of real, macroscopic systems can be adequately represented by either approach. We will also see, however, that it will often be convenient to set up general problems using the techniques of classical thermodynamics, which provide general rules describing the behavior of systems, but no specific values for the parameters of interest, then use the methods of statistical thermodynamics, which can provide specific values of parameters based either on model calculations or on nonthermodynamic (e.g., spectroscopic) measurements, to determine the values of the parameters appropriate to particular systems.

For example, we may determine using classical thermodynamics that, for an ideal gas, the heat capacity at constant pressure, C_P, is related to the heat capacity at constant volume, C_V, by

$$C_P = C_V + R, \tag{1.1}$$

where R is the gas constant, but we cannot determine, from classical thermodynamics, an absolute value for C_V without carrying out a direct experimental measurement. The use of statistical thermodynamics will enable us to determine, on the basis of model calculations and spectroscopic data, that $C_V = \frac{3}{2}R$ for a monatomic ideal gas and that $C_V \geq \frac{5}{2}R$ for an ideal gas of diatomic molecules, and it will provide means of calculating the temperature dependence of the heat capacity. We will also be able to answer, using statistical thermodynamics, such questions as why the entropy change on melting is approximately 8 J/mol-K for many metals, why the heat capacity of most simple solids approaches 50 J/mol-K at high temperatures and appears to approach zero at very low temperatures, and why the temperature dependence of the equilibrium vapor pressure over a condensed phase has the observed form.

In carrying out these treatments, we will in all cases begin by developing the general relations that can be developed from classical thermodynamics, which do not require an understanding of the system at the molecular level, then proceed, through the use of nonthermodynamic data and the use of models based on a description of the system at the molecular level, to develop relations describing the equilibrium behavior of systems with respect to such properties as phase and chemical equilibria in systems containing both gaseous and condensed phases. We will see in the process that the principal limitation of classical thermodynamics is that it requires direct experimental thermodynamic data to be of use, while the principal limitation of statistical thermodynamics is its dependence on models which may not adequately describe reality over a wide range of system conditions.

In Chapter 2, we will begin the process of comparing the two approaches by reviewing the laws on which classical thermodynamics is based, pointing out as we go the difference in approach between classical thermodynamics and statistical mechanics. Before doing this, we will define several terms that will be used throughout our treatment.

DEFINITIONS

We must first differentiate between the macroscopic and microscopic approaches to the treatment of systems:

The *macroscopic* approach treats the system as a whole. The relevant variables used in describing the state of the system are those which apply to the system as a whole — for example, temperature, pressure, volume, and composition. The number of coordinates that must be specified in order to define the state of the system is small, generally three to six.

The *microscopic* approach treats the system as a collection of minute, discrete entities — for example, atoms, molecules, electrons, or photons. The relevant variables in this approach are those that apply to the individual particles, such as velocities, momenta, positions, or vibrational and rotational frequencies. The number of coordinates that must be specified in order to define the state of the system is very large — generally six per particle. In general, these coordinates must be treated statistically to define the state of the system.

The *macroscopic* approach is the approach of classical thermodynamics. This approach leads to the formulation of a few empirical laws of great generality. The *microscopic* approach is the approach of statistical mechanics. Here the average properties of a large collection of particles are deduced using classical or quantum statistical mechanics. Note that in both cases we will be dealing with macroscopic systems: The number of particles in the system will always be large.

Continuing with definitions, we may define a *thermodynamic coordinate* as any *macroscopic* quantity having a bearing on the internal state of a system. Typical thermodynamic coordinates are the pressure and volume of a gas, the surface tension and area of a surface film, and the electromotive force and state of charge of an electrochemical cell.

These thermodynamic coordinates can be further subdivided into *intensive* and *extensive* coordinates. *Intensive* coordinates are those whose value does not change if the amount of the system is changed. Variables such as temperature, pressure, surface tension, and electromotive force are intensive coordinates.

Extensive coordinates are those whose values are directly proportional to the amount of the system present. Variables such as volume, area, and amount of charge are extensive coordinates.

A *thermodynamic system* is any entity that can be described in terms of thermodynamic coordinates. Generally one defines the system as that part of the univese that is treated in a given problem. The rest of the universe is considered surroundings. The choice of just what to include as the system in any given case is a matter of convenience. We will see examples of this later.

We may define a *thermodynamic state*, or a *state of internal equilibrium*, by saying that a system is in a given *thermodynamic state* when it meets two criteria, namely:

1. The thermodynamic coordinates of the system are either (a) uniform throughout or (b) well-defined for all volume elements of the system large enough that a macroscopic description applies.
2. The thermodynamic coordinates have no tendency to change with time.

Note that both of these conditions must be met in order for the system to be in a thermodynamic state. A typical example illustrating this definition would be a mixture of ice and water at the freezing point. The temperature and pressure would be uniform throughout the system. The density would be uniform in each phase, but would have different values in each of the two phases. Note, too, that when we say "no tendency to change with time" we exclude changes that are impossible in the situation of interest for kinetic reasons, but possible in other situations where the kinetic barrier is removed.

We can describe a thermodynamic state in two ways: We can define a *macrostate* of the system in terms of the appropriate macroscopic thermodynamic coordinates. For example, knowing the pressure, volume, and temperature of a gaseous system defines its macrostate. We can define a *microstate* of the system by specifying a microscopic description of the positions and momenta of each particle. In general, many microstates of a system will be consistent with a given macrostate. That is, if one particle in a macroscopic system changes position or if two particles exchange energy, this will result in a different microstate of the system. Microscopic changes of this sort generally do not change the macrostate of the system.

In speaking about equilibrium with respect to the above definition of a thermodynamic state, we can consider three kinds of equilibrium: *mechanical equilibrium*, which requires that there be no unbalanced mechanical forces; *thermal equilibrium*, which requires that there be no temperature gradients; and *chemical equilibrium*, which requires that there be no chemical reactions capable of proceeding at a finite rate.

The various possible thermodynamic states of a given system are related and can generally be described by an *equation of state*. This is simply a relation among the thermodynamic coordinates that specify the state of the system which is valid for a wide range of thermodynamic equilibrium states of that system. For example, the thermodynamic state of a system containing N molecules of a gas can be defined in terms of its pressure P, volume V, and temperature T. These parameters are related empirically by an appropriate equation of state, such as the ideal gas equation of state,

$$PV = NkT, \qquad (1.2)$$

where k is Boltzmann's constant, or by more sophisticated equations of state such as the van der Waals,

$$\left(P + \frac{N^2 a^2}{V^2}\right)(V - Nb) = NkT, \qquad (1.3)$$

in which a and b are constants that depend on gas composition. Similar equations in appropriate variables can be written for all thermodynamic systems. We shall see that we will be able to use statistical mechanics to develop equations of state for many systems.

If we consider now the ways in which the *thermodynamic state* of a system may change, we must introduce the concept of a *thermodynamic process*. This is any process whereby the *macroscopic thermodynamic state* of a system is changed—that is, a process in which there is a change in the measurable *thermodynamic coordinates* of the system. The occurrence of a thermodynamic process may be obvious, for example in the freezing of a liquid, or it may be quite subtle, such as in the mixing of two isotopes of the same element. In both

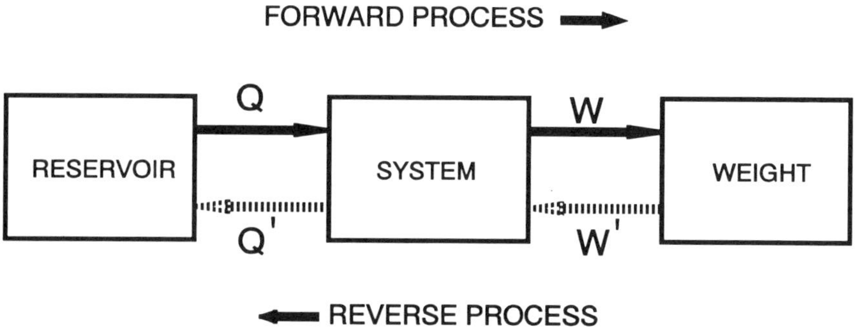

Figure 1.1 Schematic representation of a reversible process. The process is reversible if the initial and final states of the system are identical and $W = W'$, $Q = Q'$.

of these cases there is a change in one or more of the macroscopic thermodynamic coordinates. Note that a change in the microstate of the system that does not lead to a change in the macrostate of the system is *not* a thermodynamic process.

We may consider in addition some special kinds of thermodynamic processes, in which restrictions are placed on the way the process is carried out. The possible restrictions lead to the following definitions:

An *infinitesimal process* is a thermodynamic process whose extent is so small that there is only an infinitesimal change in *any* of the *thermodynamic coordinates* that describe the state of the system. For example, a dT, dV, or dP.

A *quasi-static process* is a process that is carried at a rate slow enough that the system is always very close to internal equilibrium. Note that by "slow enough" in this context we mean that the rate of the process must be slow compared to the time it takes the system to respond to the change in external conditions. For a volume or pressure change in a gas, for example, a process can be quasi-static as long as its rate is slow compared to the speed of sound in the gas.

An *adiabatic process* is one which is carried out in such a way that there is no heat flow between the system and the surroundings in the course of the process.

A *reversible process* is a process that is carried out in such a way that, at the conclusion of the process, both the system and its local surroundings may be restored to their initial state *without producing any change in the rest of the universe*. This concept of a reversible process is a very important one, so let us look at it in more detail. As an example, let us look at an idealized case, such as is shown in Figure 1.1.

A system of some sort is connected to a heat reservoir and to a weight. A process is carried out in which heat, Q, flows from the reservoir to the system, and the system causes the weight to be raised some distance, doing work, W. If, at the conclusion of the process, we can carry out the reverse process of lowering the weight to its original position and allowing the same amount of heat, Q', to flow back into the reservoir and find that the state of the system at the conclusion of the reverse process is the same as it was initially, *and* that we had no other interactions with other surroundings, then the process was reversible. If any of the above conditions are not met, then the original process was an *irreversible process*. We will find as we proceed that a *reversible process* is an idealization of any real process, but that many processes are close enough to being reversible that the reversible process is a useful concept.

The definitions that have been presented above relate to concepts that we will use throughout this treatment. We will next go on to use these definitions, along with a series of experimental observations, to develop the laws of classical thermodynamics.

BIBLIOGRAPHY

R. T. DeHoff, *Thermodynamics in Materials Science*, McGraw-Hill, New York, 1993, Chapter 2.

E. A. Guggenheim, *Thermodynamics*, North Holland, 1957, Chapters 1 and 2.

E. L. Knuth, *Introduction to Statistical Thermodynamics*, McGraw-Hill, New York, 1966, Chapter 1.

W. G. V. Rosser, *An Introduction to Statistical Physics*, Ellis Horwood, Chichester, UK, 1982, Chapter 1.

M. W. Zemansky, *Heat and Thermodynamics*, McGraw-Hill, 1957.

CHAPTER 2

THE LAWS OF CLASSICAL THERMODYNAMICS

In this chapter we will develop the laws of classical thermodynamics in terms of experimental observations that have been made in a wide variety of systems over a long period of time. We will see that this process will lead us to a small number of laws of completely general validity.

We will do this by carrying out various thermodynamic processes in various systems and under various constraints. We note first that a thermodynamic process will take place as the response of the system to a disturbance of its state of internal equilibrium arising from the interaction of the system with its surroundings. The interaction may involve a disturbance of the mechanical equilibrium of the system, by applying a force to the system which causes a mechanical displacement of the system during which work is done either by or on the system. Alternatively, the interaction may involve a disturbance in the thermal equilibrium of the system, say by impressing a temperature gradient on the system and thus causing heat flow either into or out of the system.

THE ZEROTH LAW

Let us consider first the flow of heat, or heat transfer between systems, and thermal equilibrium.

Consider two systems, separated by a common wall. We can, ideally, describe the wall as being one of two possible types: a *thermally insulating* or *adiabatic* wall, which will not permit heat flow between the two systems, or a *thermally conducting* or *diathermic* wall, which will permit heat flow between the two systems.

If we examine the process of heat flow in various systems connected by *conducting* walls, we discover, empirically, the *zeroth law of thermodynamics*, namely:

Two or more systems in mutual thermal equilibrium—that is, with no tendency for heat to flow through the conducting walls connecting them—all have the same temperature. Or, alternatively, we can say that when two systems are in thermal equilibrium with a third system, they are in thermal equilibrium with each other and all three systems have the same temperature.

WORK AND THE FIRST LAW

Consider next thermodynamic processes in which work is done either on or by a thermodynamic system. We may define *work* as a process in which a system undergoes a displacement as the result of the application of an external force. If the process that takes place is an infinitesimal process, the amount of work done may be described mathematically by

$$dW = \mathcal{F} dx, \tag{2.1}$$

where the symbol d indicates that dW is an inexact differential. That is, its value in a given process will depend on how the process is carried out. $\mathcal{F}$ and x represent, respectively, the generalized force and displacement. The exact nature of these variables will depend on the system involved. For example: For a gas at a pressure P held in a cylinder of cross-sectional area A by a piston, $\mathcal{F} = P \cdot A$, $dx = dx$, the displacement of the piston; for an electrochemical cell, $\mathcal{F} = \mathcal{E}$, the EMF of the cell, $dx = dz$, the displacement of charge; for a wire in a tensile test, $\mathcal{F} = \tau = \sigma \cdot A$, where τ is the tensile force on the wire, σ the applied stress, and A the cross-sectional area, and $dx = dl$, the change in length of the wire. By convention, the sign on the displacement is taken so as to make dW positive when work is done *by* the system. For example, $dW = P\,dV$, or $dW = -\tau\,dl$.

If we now look at processes involving the performance of work in various systems under various constraints, we discover several things:

1. In general, the amount of work done in going from an initial state of a system, for example the state represented by i on the P–V diagram in Figure 2.1, to a final state, for example f, will depend on the way in which the process was carried out. That is, the amount of work done, $W = \int P\,dV$, for the case of a gaseous system, will depend on the path in P–V space followed by the system as the process takes place.

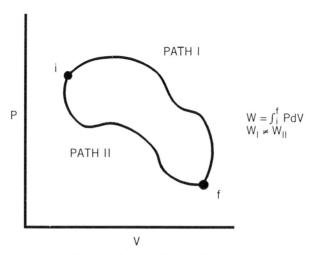

Figure 2.1 Illustration of the path dependence of the work done in passing from an initial (i) to a final (f) state.

Thus it is observed that work is not a function of the thermodynamic coordinates of the system (i.e., a thermodynamic state function) and that, as mentioned above, $đW$ is inexact.

2. If the restriction is made that the system follows a path in moving from state i to state f that involves only *adiabatic processes*, then we find that the work done,

$$W = \int_{ad} \mathscr{F}\, dx, \qquad (2.2)$$

where the subscript *ad* indicates that the integral is taken along a path involving only adiabatic processes, *is independent of the choice of path* and has the same value for all possible sequences of adiabatic paths leading from i to f. If this is so, then there must be some property of the system, which is a state function, or a function of the thermodynamic coordinates, whose value in state f differs from its value in state i by the amount $\int_{ad} \mathscr{F}\, dx$. We will call this function E, the *internal energy* of the system. Changes in the value of this function accompanying thermodynamic processes are defined by

$$E_f - E_i = -\int_{ad} \mathscr{F}\, dx. \qquad (2.3)$$

The effect of the minus sign on the right-hand side of Equation (2.3) is to make $(E_f - E_i)$ positive, and consequently the value of the integral negative, when work is done on the system. This is in accordance with the previously chosen

12   THE LAWS OF CLASSICAL THERMODYNAMICS

convention that work done on the system is negative. Note that since E is a function of the state of the system, and not of its past history, $E_f - E_i$ is independent of how we get from i to f.

3. We may now go back and look at processes that take us from i to f, irrespective of whether or not the processes are adiabatic. We know from the above discussion that $E_f - E_i$ is constant, independent of path. We also know that the only ways in which the system can interact with the environment so as to interchange energy with it, assuming the mass of the system is unchanged, are by the performance of work, W, or by the transfer of heat, Q. Consequently, we are led to the mathematical statement of the *first law of thermodynamics*.

$$E_f - E_i = Q - W, \tag{2.4}$$

or

$$dE = dQ - dW \tag{2.5}$$

for an infinitesimal process. Note that, by convention, heat flow *into* the system is taken to be positive. Finally, as a corollary to the above, if a system is taken around a cyclic process, that is, a sequence of steps that finally returns the system to its initial state, then

$$\Delta E = 0, \tag{2.6}$$

because E is a function of the state of the system only. Thus $\Delta E = 0$ for the cycle and

$$\text{net } Q \text{ in} = \text{net } W \text{ out}. \tag{2.7}$$

HEAT FLOW AND THE SECOND LAW

Now let us go on and look at processes involving heat flow under various constraints. Again we find that in general $\int dQ$ is not a constant, but is dependent on the path followed in the course of the process. That is, dQ is inexact, just like dW. However, if we restrict ourselves to processes involving only reversible paths we find something else. $\int_R dQ$, the integral of dQ along a reversible path, is still found to be a function of the path followed, but $\int_R (dQ/T)$ is found to be a constant independent of the path taken between a given set of initial and final states. In this expression, T is a property known as the *absolute temperature*, which will be defined shortly. As was the case with $\int_{ad} dW$, this implies that there is some function of the state of the system whose value at f differs from that at i by the amount $\int_R (dQ/T)$. This function is called the *entropy*, S. The change in entropy in going from i to f is defined by

$$\int_R \left(\frac{dQ}{T}\right) = S_f - S_i. \tag{2.8}$$

Consequently, we may write as the mathematical statement of the *second law of thermodynamics* that

$$S_f - S_i = \int_R \left(\frac{dQ}{T}\right), \tag{2.9}$$

or

$$dS = \frac{dQ_R}{T} \tag{2.10}$$

for an infinitesimal process. Finally,

$$dQ = T dS \tag{2.11}$$

for an infinitesimal, reversible process. Note the restriction in this definition: The entropy change $S_f - S_i$ is the same irrespective of whether the path followed between i and f is reversible or irreversible, *but* in order to calculate $S_f - S_i$ we must devise some way to transfer the system from i to f using only reversible processes, and then calculate $\int_R (dQ/T)$ for this sequence of reversible processes. In Chapter 3 we will see that consideration of entropy changes in various reversible and irreversible processes will lead us to criteria for equilibrium in a wide variety of systems.

Finally, for the special case of a reversible adiabatic process, in which $dQ = 0$, we also have $dS = 0$. Thus a reversible adiabatic process can also be referred to as an *isentropic* process.

THE ABSOLUTE TEMPERATURE

Before we leave the subject of entropy, let us develop two more concepts, those of the absolute temperature scale and the absolute zero of temperature. Since entropy is a state function, for any cyclic process the integral of dS around the full circle is

$$\oint dS = 0. \tag{2.12}$$

Let us apply this fact to a special kind of cyclic process known as a *Carnot cycle*. A Carnot cycle, shown diagrammatically in Figure 2.2, consists of four

14 THE LAWS OF CLASSICAL THERMODYNAMICS

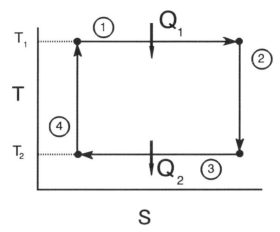

Figure 2.2 The Carnot cycle, used to define the absolute temperature scale.

steps. For the case of a gaseous system these are as follows: a reversible, isothermal expansion at a temperature T, for which

$$\Delta S_1 = \frac{Q_1}{T_1} \tag{2.13}$$

and Q_1 represents heat absorbed by the system; a reversible adiabatic expansion, for which $\Delta S = 0$; a reversible isothermal compression, for which

$$\Delta S_2 = \frac{Q_2}{T_2} \tag{2.14}$$

and Q_2 represents heat rejected by the system; and finally a reversible, adiabatic compression to return the system to its initial state, for which $\Delta S = 0$. It can be shown that this reversible cycle is the most efficient of all conceivable cycles for converting heat to work. The point of the present discussion, however, is that since for the cycle, $\Delta S = 0$, we must have

$$\Delta S_1 = -\Delta S_2, \tag{2.15}$$

and thus

$$\left|\frac{Q_1}{T_1}\right| = \left|\frac{Q_2}{T_2}\right|. \tag{2.16}$$

If we now establish an arbitrary reference temperature, say $T_3 = 273.16\,\text{K}$, we can define any other accessible temperature in terms of the ratio of heats absorbed and rejected by a Carnot cycle operating between that temperature and T_3. That is,

$$T = 273.16 \left(\frac{Q}{Q_3}\right). \tag{2.17}$$

This process also establishes the concept of an absolute zero of temperature. From application of the second law of thermodynamics to various cyclic processes, it can be shown that for all cycles, the amount of heat liberated at T_2 is always greater than zero. Consequently, in order to have $\oint dS = 0$, we must have $T_2 \geqslant 0$. In order to reduce Q_2 to zero, the isothermal compression step of the Carnot cycle would have to be carried out at $T_2 = 0$, since $Q_2 = T_2 \Delta S_2$ for this process and ΔS_2 must be finite in order for $\oint dS = 0$.

THE THIRD LAW

As shown above, the second law of thermodynamics indicates that there is a lower limit on the temperature that can be attained by any series of processes. The final question that remains is whether or not this absolute zero of temperature is attainable in practice. This question is the province of the *third law of thermodynamics*. As is the case with the other laws developed thus far, the basis of the third law is empirical observation in a wide range of systems. In this case, the empirical observations involve the measurement of temperature and entropy changes taking place as a system is brought closer to the absolute zero of temperature defined by the Carnot cycle and the second law. There are two principal empirical observations, leading to two statements of the third law. One of these observations is that if one tries to reduce the temperature toward absolute zero by a repeated series of operations, each successive operation yields a smaller ΔT, and the decrease in ΔT for each successive operation is such that it appears that $T = 0$ will *never be reached*. This is known as the *unattainability statement* of the third law. The other observation is that if ΔS is measured for an isothermal, reversible process at a series of decreasing temperatures (for example, the freezing of a liquid), it is found that as the temperature at which the process carried out approaches absolute zero, the entropy change accompanying the process approaches zero. This leads to the *Nernst–Simon statement* of the third law, namely that $\Delta S_{trans} \rightarrow 0$, as $T_{trans} \rightarrow 0$ and the entropy of both states tends to a common value at absolute zero. Note that this does *not* say that $S \rightarrow 0$ as $T \rightarrow 0$, merely that the entropy change for the process approaches zero. At this point we have no way of knowing whether the entropy itself approaches zero. This is a subject we will discuss again later, during the development of statistical thermodynamics.

16 THE LAWS OF CLASSICAL THERMODYNAMICS

BIBLIOGRAPHY

R. T. DeHoff, *Thermodynamics in Materials Science*, McGraw-Hill, New York, 1993, Chapters 2 and 3.
D. R. Gaskill, *Introduction to Metallurgical Thermodynamics*, McGraw-Hill, New York, 1973, Chapters 2, 3, and 6.
E. A. Guggenheim, *Thermodynamics*, North Holland, 1957, Chapters 1 and 2.
C. C. Kittel and H. Kroemer, *Thermal Physics*, 2nd ed., W. H. Freeman & Co., San Francisco, 1980, Chapters 2 and 8.
H. Reiss, *Methods of Thermodynamics*, Blaisdell, New York, 1965, Chapters 3 and 4.

PROBLEMS

2.1 During a quasi-static adiabatic expansion of an ideal gas the pressure at any moment is given by

$$PV^\gamma = K,$$

where γ and K are constants. Show that the work done in expanding from a state (P_i, V_i) to (P_f, V_f) is given by

$$W = \frac{P_i V_i - P_f V_f}{\gamma - 1}.$$

If $P_i = 10^5$ Pa, $V_i = 10^{-3}$ m^3, $P_f = 2 \times 10^4$ Pa, and $V_f = 3.16 \times 10^{-3}$ m^3, how many joules of work are done if $\gamma = 1.4$?

2.2 A vessel contains 6 m^3 of helium gas at 2 K and 10^4 Pa. Take the zero of internal energy of helium to be at this point.
   (a) The temperature is increased at constant volume to 288 K. Assuming helium to behave as an ideal monatomic gas, for which $dE = C_V dT$, how much heat is absorbed and what is the internal energy of the helium? Can this energy be regarded as stored heat or stored work?
   (b) The helium is now expanded adiabatically to 2 K. How much work is done? What is the new internal energy?
   (c) The helium is now compressed isothermally to its original volume. What are the quantities of heat and work in this process?

2.3 A thick-walled, insulated metal chamber contains n_i moles of helium at high pressure, P_i. It is connected through a valve to a large, almost empty gas holder in which P is maintained at $P_0 = 1$ atm. The valve is opened slightly, and helium flows slowly and adiabatically into the gas holder until the pressures on the two sides are equalized. Prove that

$$E_i - \left(\frac{n_f}{n_i}\right) E_f = \left(1 - \frac{n_f}{n_i}\right) H',$$

where

n_f = number of moles of helium left in chamber,
E_i = initial molar internal energy of helium in chamber,
E_f = final molar internal energy of helium in chamber,
$H' = E' + P_0 V'$, where E' is the molar internal energy of helium in the gas holder and V' is the molar volume of helium in the gas holder.

2.4 A vessel with rigid walls, thoroughly insulated, is divided into two parts by a partition. One part contains a gas and the other is evacuated. If the partition is suddenly broken, show that the initial and final internal energies of the gas are equal.

2.5 When a system is taken from a to b in the figure below along the path acb, 80 joules of heat flow into the system, and the system does 30 joules of work.
  (a) How much heat flows into the system along path adb, if the work done is 10 joules?
  (b) When the system is returned from b to a along the curved path, the work done on the system is 20 joules. Does the system absorb or liberate heat? If so, how much?
  (c) If $E_a = 0$ and $E_d = 40$ joules, find the heat absorbed in processes ad and db.

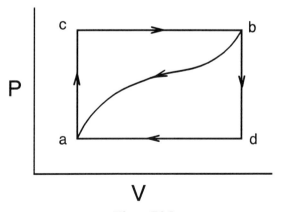

Figure P2.5

2.6 A vessel of volume V_B contains n moles of gas at high pressure. Connected to the vessel is a capillary tube through which the gas may slowly leak out into the atmosphere, where $P = P_0$. Surrounding the vessel and capillary is a water bath, in which is immersed an electric resistor. The gas is allowed to leak slowly through the capillary into the atmosphere while, at the same time, electrical energy is dissipated in the resistor at such a rate that the temperature of the gas, the vessel, the capillary, and the water is kept equal to that of the surrounding air. Show that, after as much as gas has leaked as is possible during time τ, the change in internal energy is

$$E = \varepsilon i \tau - P_0(nv_0 - V_B),$$

where

v_0 = molar volume of gas at $P = P_0$,

ε = potential difference across resistor,

i = current through the resistor.

2.7 Assume that we have a system consisting of 20 moles of an ideal gas contained in an insulated vessel at 2×10^5 Pa pressure (maintained by a weight on the top of the vessel) and at a temperature of 300 K. The vessel also contains a resistor whose resistance is $10\,\Omega$ connected to an external generator which can generate a potential of 10 volts. The generator is operated by allowing a weight to fall under the force of gravity, thus turning the generator.

(a) Determine the P, V, and T which the gas will reach if the generator is allowed to operate for 10 seconds. Assume $C_P = \frac{5}{2}R$.

(b) Verify that the work done in (a) is identical to the work done in reaching the same final P, V, and T by first slowly decreasing the weight on the top until $P = 10^5$ Pa, then operating the generator as long as is necessary, then increasing the weight until the same final conditions are reached.

2.8 The tension in a wire is increased quasi-statically and isothermally from τ_1 to τ_2. If the length, cross-sectional area, and isothermal Young's modulus (Y) of the wire remain practically constant, show that the work done is

$$W = -\frac{L}{2AY}(\tau_2^2 - \tau_1^2).$$

CHAPTER 3

CRITERIA FOR EQUILIBRIUM

We are now in a position to use the definitions and the mathematical statements of the first and second laws of thermodynamics developed in the previous chapters to develop criteria for defining states of equilibrium in chemical systems.

THE ENTROPY PRINCIPLE

To develop these criteria, we will begin by looking again at the concepts of reversibility and irreversibility. More specifically, we will look at the entropy changes that take place in the course of a reversible or an irreversible process. If we look at the entropy change accompanying any process, we can, in principle, break up the total entropy change into (a) contributions from the entropy change in the system under study and (b) the entropy change in the surroundings of the system. That is,

$$\Delta S_{univ} = \Delta S_{sys} + \Delta S_{sur}. \tag{3.1}$$

Empirical observations of ΔS_{univ} for a large number of reversible and irreversible processes in a large number of systems indicates that

$\Delta S_{univ} > 0$ for irreversible processes (also called natural or spontaneous processes),

$\Delta S_{univ} = 0$ for reversible processes,

$\Delta S_{univ} < 0$ never (a process leading to $S_{univ} < 0$ would be called an unnatural process).

This is called the *entropy principle.*

The above can also be stated in terms of infinitesimal processes as

$$dS_{univ} = dS_{sys} + dS_{sur} \qquad (3.2)$$

and

$dS_{univ} > 0$ for infinitesimal irreversible processes,

$dS_{univ} = 0$ for infinitesimal reversible processes,

$dS_{univ} < 0$ never.

We may now go on and use the mathematical statements of the first and second laws developed previously, along with the entropy principle, to develop criteria for equilibrium in chemical systems under various constraints.

DEFINITIONS

First let us make some additional definitions. A *chemical system* is any system composed of one or more chemical species, or some mixture of chemical species, in which we are primarily interested in the bulk physical and chemical properties of the system. That is, we do not have to concern ourselves with electrochemical or surface phenomena, or with stress or strain energy terms. (There is no fundamental reason for restricting ourselves to this definition. We could equally well consider these other forms of work, but will not do so at present, in order to keep the argument as simple as possible.) A *phase* is defined as any chemical system, or any physically distinguishable portion of a chemical system, which is in a well-defined state of internal equilibrium. For example, in a chemical system composed of iron and carbon atoms, the equilibrium phases might be ferrite, austenite, or cementite.

Phases can be subdivided into *closed phases*, in which there is no change in the amount of material present in the course of a process taking place, and *open phases*, in which changes in the amount of material present are permitted in the course of the process.

A *component* of a chemical system is simply any one of the independently variable chemical species that make up the phase—for example, iron and carbon in the example cited above. The restriction implied in the phrase "independently variable" is an important one. For example, in a mixture of NaCl in H_2O, the components are NaCl and H_2O, not Na, Cl, H, and O.

For completeness we will also define a *constituent* as any observable structural feature of a material—for example, pearlite in the iron–carbon system used in the example above. Constituent is *not* a thermodynamic term and will not be used further in this treatment.

THE BASIC EQUILIBRIUM POSTULATE

To develop our criteria for equilibrium, we must make an additional postulate that arises as a consequence of our definitions of thermodynamic equilibrium and of reversible processes, namely:

In a system that is in internal equilibrium, any infinitesimal process about a point of equilibrium is reversible.

Note that this is a postulate—that is, a statement whose validity rests on the fact that the behavior of real systems is in all cases consistent with the statement. The rationale for this postulate can be seen if we recall the definition of reversibility. If we are considering processes that take place only infinitesimally removed from a point of equilibrium and require only an infinitesimal change in conditions to cause reversal, the likelihood of introducing dissipative processes is minimized.

Keeping this postulate in mind, we may write, for a one-phase, closed chemical system (which may contain one or more components) in internal equilibrium

$$dE = dQ - dW \tag{3.3}$$

for any infinitesimal process. Since the system is initially in equilibrium, the process, according to the postulate just made, will be reversible. From our statement of the second law, we know that for an infinitesimal, reversible process we obtain

$$dQ = TdS. \tag{3.4}$$

We also know that from the way we have defined a chemical system the only type of work possible involves a change in volume due to an applied pressure, or

$$dW = PdV. \tag{3.5}$$

Thus

$$dE = TdS - PdV \tag{3.6}$$

for a closed phase in internal equilibrium. In order to generalize this to the case of a one-phase, open chemical system of one or more components, we can define

$$\left(\frac{\partial E}{\partial N_i}\right)_{S,V,N_{j \neq i}} \equiv \mu_i, \tag{3.7}$$

the chemical potential of component i, in which N is the number of molecules of component i. This yields

$$dE = TdS - PdV + \sum_i \mu_i dN_i \tag{3.8}$$

for a one-phase, open chemical system in internal equilibrium.

ADDITIONAL STATE FUNCTIONS

Before proceeding any further along this line, we will define some additional thermodynamic functions:

$$H \equiv E + PV, \tag{3.9}$$

the enthalpy,

$$F \equiv E - TS, \tag{3.10}$$

the Helmholz free energy, and

$$G \equiv E + PV - TS, \tag{3.11}$$

the Gibbs free energy. Since all of these functions are defined in terms of P, V, E, T, and S, all of which are state functions, H, F, and G are also state functions. Moreover, since all of them contain E and two of them contain S, they are all indefinite in absolute value. That is, we can write expressions for the change in H, F, or G in a process but we cannot, on the basis of what we have developed thus far, write an expression for the absolute values of any of these functions.

We can, however, write the following expressions for the changes in these functions associated with an infinitesimal process about a point of equilibrium in a one-phase chemical system:

$$dE = TdS - PdV + \sum_i \mu_i dN_i, \tag{3.12}$$

$$dH = dE + PdV + VdP = TdS + VdP + \sum_i \mu_i dN_i, \tag{3.13}$$

$$dF = dE - TdS - SdT = -SdT - PdV + \sum_i \mu_i dN_i, \tag{3.14}$$

$$dG = dH - TdS - SdT = -SdT + VdP + \sum_i \mu_i dN_i. \tag{3.15}$$

Note that in this process we observe that the change in each of the functions can be expressed in terms of changes in system composition plus two of the other state functions. Thus E is said to be the *characteristic function* of the variables S and V, and similarly H of S and P, F of T and V, and G of T and P. We may also note in passing that in our initial definition of a chemical system we restricted ourselves to systems in which the only work possible was $P\,dV$. If other work terms are important in a given system, the expression for dE must be modified accordingly. For example, in a system in which the work involved in increasing the surface area of the system was significant, the expression for the internal energy would be written as follows:

$$dE = T\,dS - P\,dV + \gamma\,dA + \sum_i \mu_i\,dN_i. \tag{3.16}$$

This will, of course, also modify the expressions for dH, dF, and dG. We will see examples of this later.

We may also note in passing, without going through the detailed proof, that the expression for dE can be integrated to give

$$E = TS - PV + \sum_i \mu_i N_i. \tag{3.17}$$

This expression for E is still undetermined in absolute value to the extent of arbitrary additive constants in S and the μ_i.

We may also consider substitution of the integrated expression for E into the defining relation for G,

$$G \equiv E + PV - TS, \tag{3.18}$$

yielding

$$G = TS - PV + \sum_i \mu_i N_i + PV - TS \tag{3.19}$$

or

$$G = \sum_i \mu_i N_i. \tag{3.20}$$

That is, μ_i is simply the contribution per molecule to the Gibbs free energy of the ith component of the system.

CRITERIA FOR EQUILIBRIUM

We are now in a position to go on and use the relations developed above to set up criteria for equilibrium in chemical systems. Recall that the entropy principle could be stated as

$$\Delta S_{univ} = \Delta S_{sys} + \Delta S_{sur} \geq 0, \tag{3.21}$$

depending on whether the process was irreversible (> 0) or reversible ($= 0$). Or, for an infinitesimal process

$$dS_{univ} = dS_{sys} + dS_{sur} \geq 0. \tag{3.22}$$

If we assume that in an infinitesimal process dS_{sur} is brought about by reversibly transferring heat dQ to or from the system, and we can always do this in principle, then

$$dS_{sur} = -\frac{dQ}{T}, \tag{3.23}$$

where a positive dQ implies heat transfer to the system. Thus

$$dS_{univ} = -\frac{dQ}{T} + dS_{sys} \geq 0, \tag{3.24}$$

or, upon rearranging,

$$dQ - TdS_{sys} \leq 0. \tag{3.25}$$

Moreover, we know that for a chemical system

$$dQ = dE_{sys} + PdV_{sys}. \tag{3.26}$$

Making this substitution, we obtain

$$dE_{sys} + PdV_{sys} - TdS_{sys} \leq 0. \tag{3.27}$$

That is, we now have a criterion for reversibility and, hence, through the postulate stated earlier, for equilibrium, based only on changes taking place *in the system* in the course of the infinitesimal process. We no longer need to measure dS_{univ} to determine whether the process was reversible or irreversible.

THE ISOLATED SYSTEM

We may now apply this general relation to a wide range of systems, in situations where the values of one or more of the thermodynamic coordinates are restricted, to develop criteria for equilibrium in the system in terms of changes in only one of the state functions. For example, if we have an infinitesimal process taking place in a system of a fixed number of particles, N, at fixed total volume, V, surrounded by insulating walls—that is, a system of constant N, V, and E, or what we will refer to as an *isolated system*—we may write

$$dE + PdV - TdS \leqslant 0, \tag{3.28}$$

or, since in this case $dE = 0$, $dV = 0$, we have

$$dS \geqslant 0. \tag{3.29}$$

That is, if we carry out an infinitesimal process in a system of constant N, V, and E and determine that $dS_{sys} = 0$ for that process, then the process was *reversible* and, according to the postulate stated previously, the system was in equilibrium before the process began.

Another way of looking at this is to say that in a system of fixed N, V, and E, the entropy will be a maximum at equilibrium.

THE CLOSED ISOTHERMAL SYSTEM

We may deduce similar criteria for any other set of constraints. We will look explicitly at three other common cases. First, consider the case of a system with a constant number of particles, N, and a fixed total volume, V, placed in contact with a heat reservoir at a constant temperature, T. This is commonly referred to as a *closed isothermal system*. As before, we may write

$$dE + PdV - TdS \leqslant 0. \tag{3.30}$$

However, for this case

$$d(TS) = TdS + SdT = TdS \tag{3.31}$$

because T is constant. Thus

$$dE - d(TS) \leqslant 0, \tag{3.32}$$

or since

$$F = E - TS$$

we have

$$dF \leqslant 0. \tag{3.33}$$

Thus, in a closed isothermal system, F is a *minimum* at equilibrium.

THE CLOSED ISOBARIC SYSTEM

As a third case, we may consider a system made up of a fixed number of particles, N, at a constant total pressure, P, in contact with a reservoir at temperature, T. That is, a system of fixed N, P, and T. We begin again with

$$dE + PdV - TdS \leqslant 0. \tag{3.34}$$

For this case,

$$d(TS) = TdS + SdT = TdS, \tag{3.35}$$

because T is constant,

$$d(PV) = PdV + VdP = PdV \tag{3.36}$$

as P is constant.
Thus

$$dE + d(PV) - d(TS) \leqslant 0 \tag{3.37}$$

for this case, or since

$$G = E + PV - TS$$

we have

$$dG \leqslant 0. \tag{3.38}$$

Thus, in a system of constant N, P, and T the Gibbs free energy is a minimum at equilibrium.

THE OPEN ISOTHERMAL SYSTEM

Finally, we will look at the case of an open system, in which the amount of material may change in a process, but in such a way that the chemical potential, μ_i, of each species remains constant. We will also fix the volume, V, and the temperature, T. Such a system of fixed μ_i, V, T is referred to as an *open isothermal system*. To develop the criterion for equilibrium in this case, we must use a form of our basic equation that allows for a change in the amount of material in the system during a process. The required equation is:

$$dE + PdV - TdS - \sum_i \mu_i dN_i \leqslant 0. \tag{3.39}$$

For the case of a system of constant μ_i, V, and T we have

$$PdV = 0, \qquad (3.40)$$

and

$$dG = d\left(\sum_i \mu_i N_i\right) = \sum_i \mu_i dN_i + \sum_i N_i d\mu_i = \sum_i \mu_i dN_i \qquad (3.41)$$

at constant μ_i. Moreover, since T is constant, $TdS = d(TS)$ and

$$dE - d(TS) - dG \leqslant 0. \qquad (3.42)$$

Thus, since

$$G = E - TS + PV$$

we have

$$dE - d(TS) - dE + d(TS) - d(PV) \leqslant 0, \qquad (3.43)$$

or

$$d(PV) \geqslant 0. \qquad (3.44)$$

In other words, in an open isothermal system having constant μ_i, V, and T, at equilibrium, PV tends to a maximum.

Thus we see that a criterion for equilibrium in systems whose state is characterized by three of the macroscopic thermodynamic state functions can be stated in terms of a maximum of one, or at most two, other thermodynamic state functions. Note that although the general criterion for equilibrium was formulated in terms of a maximum in the entropy of the *universe*, a maximum in the entropy of the *system* is the criterion for equilibrium only in an *isolated system*. We will see, when we develop our statistical thermodynamic framework, that the functions S, F, and PV have special significance in the description of isolated, closed isothermal and open isothermal systems, respectively.

BIBLIOGRAPHY

D. R. Gaskill, *Introduction to Metallurgical Thermodynamics*, McGraw-Hill, New York, 1973, Chapters 3 and 5.

E. A. Guggenheim, *Thermodynamics*, North Holland, 1957, Chapters 1 and 2.

C. C. Kittel and H. Kroemer, *Thermal Physics*, 2nd ed., W. H. Freeman & Co., San Francisco, 1980, Chapters 2 and 5.

E. L. Knuth, *Introduction to Statistical Thermodynamics*, McGraw-Hill, New York, 1966, Appendix 3.

H. Reiss, *Methods of Thermodynamics*, Blaisdell, New York, 1965, Chapters 3 and 4.

PROBLEMS

3.1 Determine the criterion for equilibrium in a system of constant H, P, and N.

3.2 Determine the criterion for equilibrium in an open isobaric system (i.e., a system of constant P, V, and μ).

3.3 The figure below shows a thermally insulated chamber, containing an ideal gas. Within the chamber is a paddle wheel, connected externally to a weight which can be lowered to operate the paddle wheel. Prove thermodynamically that the process of turning the paddle wheel by lowering the weight is irreversible.

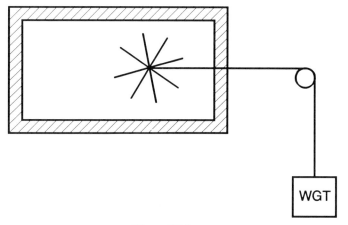

Figure P3.3

3.4 **(a)** One kilogram of water at 273 K is brought into contact with a heat reservoir at 373 K. When the water has reached 373 K, what is the entropy change of the water? Of the heat reservoir? Of the universe?

(b) If the water had been heated from 273 to 373 K by first bringing it in contact with a reservoir at 323 K and then with a reservoir at 373 K, what would have been the entropy change of the universe?

CHAPTER 4

USEFUL MATHEMATICAL RELATIONS

We will now spend some time developing a series of relationships among the various thermodynamic state functions. These relations will prove useful in the solution of many thermodynamic problems and will provide a means of expressing changes in the thermodynamic state functions in terms of readily measurable experimental parameters. We will do this explicitly for the case of a *one-phase, closed chemical system*, but keeping in mind that we could extend it to systems of more than one phase or to open phases if we so desired.

PARTIAL DERIVATIVES

We have already determined that

$$dE = TdS - PdV, \tag{4.1}$$

$$dH = TdS + VdP, \tag{4.2}$$

$$dF = -SdT - PdV, \tag{4.3}$$

and

$$dG = -SdT + VdP. \tag{4.4}$$

We can state, in addition, that the thermodynamic state of a closed, one-phase chemical system can be specified by specifying the values of *any two* of the thermodynamic coordinates of the system. That is, for a given number of molecules, N, in a one-component system (or $N_1, N_2 \cdots$ in a multicomponent system) the state of the system is a function of the N_i and any two other thermodynamic coordinates. For example, we could express the state of the system as a function of N, V, E, or of N, V, T, or of N, S, H, or of any other two arbitrarily chosen thermodynamic coordinates.

Since the state of a closed system can be specified by specifying any two of the thermodynamic coordinates, *changes* in any of the thermodynamic coordinates accompanying any thermodynamic process in that system can be expressed in terms of the *changes* in any two of the other thermodynamic coordinates. For example, we could express the change in E in a process as

$$dE = \left(\frac{\partial E}{\partial T}\right)_V dT + \left(\frac{\partial E}{\partial V}\right)_T dV \tag{4.5}$$

or as

$$dE = \left(\frac{\partial E}{\partial P}\right)_H dP + \left(\frac{\partial E}{\partial H}\right)_P dH, \tag{4.6}$$

or similarly for any other arbitrarily chosen set of one dependent and two independent variables.

In order to use any of the general relations of the type shown above, one must be able to evaluate the partial derivatives that appear in the expressions in terms of the physically measurable parameters of the system. For some choices of independent variables, this is straightforward. For example, if we consider the case of $E = f(S, V)$, the general expression is

$$dE = \left(\frac{\partial E}{\partial S}\right)_V dS + \left(\frac{\partial E}{\partial V}\right)_S dV. \tag{4.7}$$

This may be compared to the previously developed expression

$$dE = TdS - PdV \tag{4.8}$$

to yield

$$\left(\frac{\partial E}{\partial S}\right)_V = T, \quad \left(\frac{\partial E}{\partial V}\right)_S = -P. \tag{4.9}$$

Similar expressions could also be written using the previously developed expression for dH, dF, and dG.

RELATION TO MEASURABLE PROPERTIES

It is also helpful in the evaluation of the partial derivatives in these general equations to define certain partial derivatives in terms of the measurable properties of the system. For example, we may define the following:

The coefficient of volume thermal expansion, α

$$\alpha \equiv \frac{1}{V}\left(\frac{\partial V}{\partial T}\right)_P. \qquad (4.10)$$

The isothermal compressibility, κ

$$\kappa \equiv -\frac{1}{V}\left(\frac{\partial V}{\partial P}\right)_T. \qquad (4.11)$$

The heat capacity at constant volume, C_V

$$C_V \equiv \left(\frac{\partial Q}{\partial T}\right)_V = \left(\frac{\partial E}{\partial T}\right)_V \qquad (4.12)$$

(because, from the first law $dE = dQ - PdV = dQ$ if $dV = 0$).
The heat capacity at constant pressure, C_P

$$C_P \equiv \left(\frac{\partial Q}{\partial T}\right)_P = \left(\frac{\partial H}{\partial T}\right)_P \qquad (4.13)$$

(because $dH = dE + d(PV) = dQ + VdP = dQ$ if $dP = 0$).

EVALUATION OF PARTIAL DERIVATIVES

With these relations in mind, let us look at how we can evaluate the general partial derivatives in a particular case. As an example, let us develop an expression for $S = f(V, T)$. We can write the general relation

$$dS = \left(\frac{\partial S}{\partial T}\right)_V dT + \left(\frac{\partial S}{\partial V}\right)_T dV. \qquad (4.14)$$

This may be compared with the previously developed expression

$$dE = TdS - PdV, \qquad (4.15)$$

which can be arranged to give

$$dS = \frac{1}{T}(dE + PdV) \qquad (4.16)$$

or, expressing dE as a $f(T, V)$,

$$dS = \frac{1}{T}\left[\left(\frac{\partial E}{\partial T}\right)_V dT + \left(\frac{\partial E}{\partial V}\right)_T dV + P dV\right], \quad (4.17)$$

or

$$dS = \frac{1}{T}\left(\frac{\partial E}{\partial T}\right)_V dT + \frac{1}{T}\left(P + \left(\frac{\partial E}{\partial V}\right)_T\right) dV. \quad (4.18)$$

Comparison of the two expressions for dS [equations (4.14) and (4.18)] yields

$$\left(\frac{\partial S}{\partial T}\right)_V = \frac{1}{T}\left(\frac{\partial E}{\partial T}\right)_V = \frac{C_V}{T}, \quad \left(\frac{\partial S}{\partial V}\right)_T = \frac{1}{T}\left[P + \left(\frac{\partial E}{\partial V}\right)_T\right]. \quad (4.19)$$

THE CROSS-DIFFERENTIATION IDENTITY

In order to carry this development any further, we must make use of another general mathematical relation, the cross differentiation identity. This relation states that

$$\frac{\partial}{\partial x}\left(\frac{\partial W}{\partial y}\right) = \frac{\partial}{\partial y}\left(\frac{\partial W}{\partial x}\right), \quad (4.20)$$

or

$$\left(\frac{\partial^2 W}{\partial x \partial y}\right) = \left(\frac{\partial^2 W}{\partial y \partial x}\right), \quad (4.21)$$

where W is any continuous function of x and y, and the first partial derivatives are also continuous. These criteria will always be met in the real physical systems treated by thermodynamics.

We can demonstrate the utility of this relation by applying it to the determination of $S = f(T, V)$ that we are carrying out. It was shown above that

$$\left(\frac{\partial S}{\partial T}\right)_V = \frac{1}{T}\left(\frac{\partial E}{\partial T}\right)_V, \quad \left(\frac{\partial S}{\partial V}\right)_T = \frac{1}{T}\left[P + \left(\frac{\partial E}{\partial V}\right)_T\right]. \quad (4.22)$$

Differentiation of the first of these relations by dV and the second by dT yields

$$\left(\frac{\partial^2 S}{\partial V \partial T}\right) = \frac{1}{T}\left(\frac{\partial^2 E}{\partial V \partial T}\right) \quad (4.23)$$

and

$$\left(\frac{\partial^2 S}{\partial T \partial V}\right) = \frac{1}{T}\left[\left(\frac{\partial P}{\partial T}\right)_V + \left(\frac{\partial^2 E}{\partial T \partial V}\right)\right] - \frac{1}{T^2}\left[P + \left(\frac{\partial E}{\partial V}\right)_T\right]. \quad (4.24)$$

Using the cross-differentiation identity, these two expressions must be equal. Thus,

$$\frac{1}{T}\left(\frac{\partial^2 E}{\partial V \partial T}\right) = \frac{1}{T}\left[\left(\frac{\partial P}{\partial T}\right)_V + \left(\frac{\partial^2 E}{\partial T \partial V}\right)\right] - \frac{1}{T^2}\left[P + \left(\frac{\partial E}{\partial V}\right)_T\right], \quad (4.25)$$

or since

$$\left(\frac{\partial^2 E}{\partial V \partial T}\right) = \left(\frac{\partial^2 E}{\partial T \partial V}\right) \quad (4.26)$$

we have

$$\left(\frac{\partial P}{\partial T}\right)_V - \frac{1}{T}\left(\frac{\partial E}{\partial V}\right)_T - \frac{P}{T} = 0, \quad (4.27)$$

or

$$\left(\frac{\partial E}{\partial V}\right)_T + P = T\left(\frac{\partial P}{\partial T}\right)_V. \quad (4.28)$$

Or, since

$$\left(\frac{\partial P}{\partial T}\right)_V = \frac{\frac{1}{V}\left(\frac{\partial V}{\partial T}\right)_P}{-\frac{1}{V}\left(\frac{\partial V}{\partial P}\right)_T} = \frac{\alpha}{\kappa}, \quad (4.29)$$

we have

$$\left(\frac{\partial E}{\partial V}\right)_T + P = T\left(\frac{\alpha}{\kappa}\right). \quad (4.30)$$

This may be substituted into the previously developed relation for $S = f(V, T)$,

$$dS = \frac{1}{T}\left(\frac{\partial E}{\partial T}\right)_V dT + \frac{1}{T}\left[\left(\frac{\partial E}{\partial V}\right)_T + P\right]dV, \quad (4.31)$$

to yield

$$dS = \left(\frac{C_V}{T}\right) dT + \left(\frac{\alpha}{\kappa}\right) dV, \qquad (4.32)$$

the desired relation for $S = f(T, V)$ in terms of observables.

MAXWELL'S RELATIONS

It is also useful to apply the cross-differentiation identity to the expressions for dE, dH, and so on, developed earlier. For example,

$$dE = TdS - PdV, \qquad (4.33)$$

or, in general,

$$dE = \left(\frac{\partial E}{\partial S}\right)_V dS + \left(\frac{\partial E}{\partial V}\right)_S dV. \qquad (4.34)$$

Thus

$$\left(\frac{\partial E}{\partial S}\right)_V = T, \qquad \left(\frac{\partial E}{\partial V}\right)_S = -P, \qquad (4.35)$$

or, applying the cross-differentiation identity we obtain

$$\left(\frac{\partial^2 E}{\partial V \partial S}\right) = \left(\frac{\partial T}{\partial V}\right)_S, \qquad \left(\frac{\partial^2 E}{\partial S \partial V}\right) = -\left(\frac{\partial P}{\partial S}\right)_V. \qquad (4.36)$$

Thus

$$\left(\frac{\partial T}{\partial V}\right)_S = -\left(\frac{\partial P}{\partial S}\right)_V, \quad \text{or} \quad \left(\frac{\partial S}{\partial P}\right)_V = -\left(\frac{\partial V}{\partial T}\right)_S. \qquad (4.37)$$

One can proceed in the same fashion with the other three similar expressions, to obtain

$$\left(\frac{\partial T}{\partial P}\right)_S = \left(\frac{\partial V}{\partial S}\right)_P, \quad \text{or} \quad \left(\frac{\partial S}{\partial V}\right)_P = \left(\frac{\partial P}{\partial T}\right)_S, \qquad (4.38)$$

$$\left(\frac{\partial S}{\partial V}\right)_T = \left(\frac{\partial P}{\partial T}\right)_V, \qquad (4.39)$$

$$\left(\frac{\partial S}{\partial P}\right)_T = -\left(\frac{\partial V}{\partial T}\right)_P. \qquad (4.40)$$

This set of relations is known as *Maxwell's relations for thermodynamics*. The utility of these relations is that they provide a means of determining changes in S by knowing the relations among P, V, and T, which can be expressed in terms of α, κ, C_V, and C_P.

We now have the means at hand to set up an expression for the differential of any of the state functions of a given system in terms of the differentials of any two other state functions and coefficients involving only state functions and their measurable partial derivatives. We do this by using the four general equations derived from the first and second laws, the four definitions of α, κ, C_V, and C_P, the cross-differentiation identity, and the four Maxwell's equations. These are sufficient for any case of practical interest.

INTEGRATION OF THE STATE FUNCTION EQUATIONS

Finally, we can go one step farther. Since only state functions and their derivatives are involved in any of the equations developed using the techniques discussed above, the resulting equations can be integrated over appropriate limits to determine the finite change in one state function over the course of a process that is finite in extent. For example, we showed that

$$dS = \left(\frac{C_V}{T}\right) dT + \left(\frac{\alpha}{\kappa}\right) dV. \tag{4.41}$$

For a process involving the change of the state of the system from T_i, V_i to T_f, V_f, we may write

$$\Delta S = \int dS = \int_{T_i}^{T_f} \left(\frac{C_V}{T}\right) dT + \int_{V_i}^{V_f} \left(\frac{\alpha}{\kappa}\right) dV. \tag{4.42}$$

In evaluating the integrals it is necessary to keep in mind that, in general, $C_V = f(T)$ and $\alpha, \kappa = f(V)$ and that this dependence must be accounted for in the evaluation of the integral.

BIBLIOGRAPHY

R. T. DeHoff, *Thermodynamics in Materials Science*, McGraw-Hill, New York, 1993, Chapter 4.

D. R. Gaskill, *Introduction to Metallurgical Thermodynamics*, McGraw-Hill, New York, 1973, Chapter 5.

E. A. Guggenheim, *Thermodynamics*, North Holland, 1957, Chapter 3.

H. Reiss, *Methods of Thermodynamics*, Blaisdell, New York, 1965, Chapter 2.

J. W. Whalen, *Molecular Thermodynamics*, John Wiley & Sons, New York, 1991, Appendix B.

PROBLEMS

4.1 Express E as a function of T and P for a closed system.

4.2 Show that

$$\left(\frac{\partial T}{\partial S}\right)_V = \frac{T}{C_V} \quad \text{and} \quad \left(\frac{\partial T}{\partial S}\right)_P = \frac{T}{C_P}.$$

4.3 Define V, S, E, H, and μ_i in terms of G, P, T, and partial derivatives of these coordinates.

4.4 One mole of benzene is cooled isobarically and reversibly from 30°C to 20°C at a pressure of 10^5 Pa.
 (a) How much heat must be extracted to do this?
 (b) How much work is done by the benzene in this process?

4.5 A nickel wire $1\,\text{mm}^2$ in cross-sectional area and 100 cm long is stretched uniformly in tension — elastically, isothermally, and reversibly — at 300 K from its rest position to a stress of $10^5\,\text{dyn/cm}^2$ and then cut.
 (a) Draw a diagram of the processes occurring on a τ–l plot.
 (b) Calculate the work done, the heat flow into the wire, and the wire's change in internal energy during the stressing process.
 (c) When the wire is cut, what is its change in temperature?
 (d) What is the entropy change of the universe associated with this irreversible process?

4.6 A soap film is formed by stretching $1\,\text{cm}^3$ of a soap solution, slowly and isothermally at 300 K, on a frame as shown in the figure below.

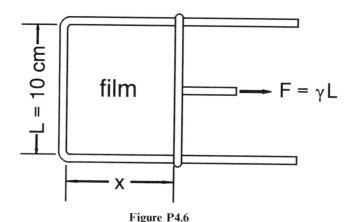

Figure P4.6

(a) How much work is done in extending the film from $x = 0$ to $x = 10$ cm?

(b) What is the entropy change associated with the stretching process?

(c) If the film breaks and the solution forms a drop having negligible surface area, what is the final temperature of the drop?

For water, $\gamma = 8 \times 10^{-6}$ J/cm², $C_P = 4.18$ J/mol-K, $(d\gamma/dT)_A = -2 \times 10^{-12}$ J/cm²-K, and $\rho = 1$ g/cm³.

4.7 One kilogram of water is compressed reversibly and isothermally at 300 K from 10^5 Pa to 2×10^6 Pa. Calculate the work done on the water, the heat flow into the water and its increase in internal energy. The isothermal compressibility of water is 5×10^{-10} Pa^{-1}, and its volume thermal expansion coefficient is 2×10^{-4} K^{-1}.

CHAPTER 5

GENERAL THEORY OF PHASE STABILITY

We will return to our discussion of criteria for equilibrium in chemical systems and then go on to show how the criteria developed previously for a closed, single-phase chemical system can be extended to characterize equilibrium in systems containing more than one phase, and in which the phases may be open.

GENERAL RELATIONS

We begin by recalling the relations developed for changes in the thermodynamic state of a one-phase open system:

$$dE = TdS - PdV + \sum_i \mu_i dN_i, \tag{5.1}$$

$$dH = TdS + VdP + \sum_i \mu_i dN_i, \tag{5.2}$$

$$dF = -SdT - PdV + \sum_i \mu_i dN_i, \tag{5.3}$$

$$dG = -SdT + VdP + \sum_i \mu_i dN_i. \tag{5.4}$$

Recall, too, that we stated that the expression for dE could be integrated to yield

$$E = TS - PV + \sum_i \mu_i N_i \tag{5.5}$$

40 GENERAL THEORY OF PHASE STABILITY

and that from this and the definition of G we determined that

$$G = \sum_i \mu_i N_i. \qquad (5.6)$$

Recall finally that we determined, using the first and second law equations plus the entropy principle, that the criteria for reversible (i.e., equilibrium) and irreversible changes in systems under various constraints could be written as

$dS \geq 0$ for a system of constant N, V, and E,

$dF \leq 0$ for a system of constant N, V, and T,

$dG \leq 0$ for a system of constant N, P, and T,

$d(PV) \geq 0$ for a system of constant μ, V, and T.

DEGREES OF FREEDOM

At a first step in the application of these relations to multiphase, open-phase systems, we will introduce the concept of a degree of freedom. This may be defined by stating that the number of *degrees of freedom* a system possesses is equal to the number of *independently variable, intensive properties* of the system. For the case of a macroscopic description of a chemical system, which we are using here, the appropriate intensive properties are pressure, temperature, and the chemical potentials of the various components that make up the system.

For the case of a single-phase chemical system in internal equilibrium, in which the only work that can be done is PdV, the total number of intensive variables that are involved in the description of the system is $C + 2$, where C is the number of components in the system. However, not all of these variables are independent, as we will now demonstrate. There is an equation relating these $C + 2$ variables, so that only $C + 1$ of them are *independent* and thus constitute degrees of freedom.

THE GIBBS–DUHEM EQUATION

The equation in question is called the *Gibbs–Duhem Equation*. To develop this equation, we begin with the general relation

$$G = \sum_i \mu_i N_i. \qquad (5.7)$$

This may be differentiated to yield

$$dG = \sum_i \mu_i dN_i + \sum_i N_i d\mu_i \qquad (5.8)$$

as the change in G to be expected in any infinitesimal process. An equally valid formulation for dG is the expression developed earlier from the expression for dE and the definition of G, namely,

$$dG = -SdT + VdP + \sum_i \mu_i dN_i. \tag{5.9}$$

Since both expressions are equally valid, they may be equated to yield

$$\sum_i \mu_i dN_i + \sum_i N_i d\mu_i = -SdT + VdP + \sum_i \mu_i dN_i, \tag{5.10}$$

or

$$SdT - VdP + \sum_i N_i d\mu_i = 0, \tag{5.11}$$

the desired relation among the intensive variables T, P, and μ_i that must hold for any infinitesimal process about a point of equilibrium in a single phase of a chemical system. We will return to this equation later to develop the *Gibbs Phase Rule*.

EQUILIBRIUM IN MULTIPHASE SYSTEMS

Before developing the phase rule, however, let us extend our reasoning from consideration of a single phase in internal equilibrium to the case of a system of two or more phases, each of which is in internal equilibrium and all of which are in mutual external equilibrium.

We will use as a model a system of two or more phases—for example, α and β—in contact with one another, in a container, as shown in Figure 5.1. For the system as a whole, we can write the same type of relations among the thermodynamic coordinates as we have written before. For example, for any process in this multiphase system we obtain

$$dE_{sys} = TdS_{sys} - PdV_{sys} + \sum_i \mu_i dN_{i_{sys}}, \tag{5.12}$$

and similarly for dH_{sys}, dF_{sys}, and dG_{sys}. Moreover, because E, H, F, and G are *extensive* properties we can write that

$$\Delta E_{sys} = \Delta E^\alpha + \Delta E^\beta + \cdots + \Delta E^n \tag{5.13}$$

and

$$dE_{sys} = dE^\alpha + dE^\beta + \cdots + dE^n \tag{5.14}$$

$$= \sum_j T^j dS^j - \sum_j P^j dV^j + \sum_j \sum_i \mu_i^j dN_i^j, \tag{5.15}$$

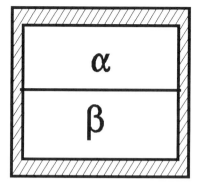

Figure 5.1 Illustration of a two-phase system. At equilibrium, the properties of the α phase will be related to those of the β phase through relations developed in this chapter.

where the summation over j is taken over all phases present, and similarly for dH, dF, and dG, where the only restriction made at this point is the absence of chemical reactions.

Let us look now at what further restrictions are placed on the above relations if we constrain the system to be at equilibrium.

THERMAL EQUILIBRIUM

If we consider first thermal equilibrium, we know from the zeroth law that all phases in mutual thermal contact, or in thermal contact through a third phase, must have the same temperature in order to be in thermal equilibrium. Thus if the system as a whole is in thermal equilibrium we obain

$$T^{\alpha} = T^{\beta} = T^{\gamma} \cdots = T. \tag{5.16}$$

HYDROSTATIC EQUILIBRIUM

Consider next the case of hydrostatic equilibrium. Here we consider a system of several phases in equilibrium at constant temperature. We let the volume of one phase (say, the α phase) increase by a volume element, dV^{α}, and at the same time let the volume of a second phase (say, the β phase) decrease by the same amount, maintaining the total volume of the system constant and maintaining the composition of each phase constant. Since we are thus dealing with a system of constant total N, V, T, the criterion for equilibrium is that, for the process involving changes in the volumes of the α and β phases,

$$dF = dF^{\alpha} + dF^{\beta} = 0. \tag{5.17}$$

Since the temperature is constant and there are no changes in composition we obtain

$$dF^\alpha = -P^\alpha dV^\alpha, \tag{5.18}$$

$$dF^\beta = -P^\beta dV^\beta = P^\beta dV^\alpha. \tag{5.19}$$

(Because V_{sys} is constant, $dV^\alpha = -dV^\beta$.) Thus

$$-P^\alpha dV^\alpha + P^\beta dV^\alpha = 0 \tag{5.20}$$

or

$$P^\alpha = P^\beta. \tag{5.21}$$

Consequently, any two phases in hydrostatic equilibrium have the same pressure.

By a similar argument, we can show that if $P^\alpha \neq P^\beta$ in a system at constant temperature, then the system is *not* in hydrostataic equilibrium. Consequently, any process in which the relative volumes of two phases change will be irreversible, and we will have

$$dF = dF^\alpha + dF^\beta < 0. \tag{5.22}$$

If we again consider the process in which the volume of the α phase increases by dV^α, and the volume of the β phase decreases by the same amount, we have

$$dF^\alpha + dF^\beta < 0, \tag{5.23}$$

$$-P^\alpha dV^\alpha + P^\beta dV^\alpha < 0, \tag{5.24}$$

or

$$P^\alpha > P^\beta. \tag{5.25}$$

That is, the phase in which the pressure is higher will increase in volume at the expense of the phase in which the pressure is lower. Thus, we see that we can use the previously developed criteria for equilibrium and for natural processes to determine both the *point of equilibrium* in a system and the *direction* of any change required to reach equilibrium. We cannot, however, say anything about the rate at which the equilibrium point will be approached.

DISTRIBUTIVE EQUILIBRIUM

Next let us consider another type of equilibrium, namely, equilibrium with respect to the distribution of the various components in the system among the phases present. In this case, we consider the transfer of a small amount of one component, i, say dN_i, from the α phase to the β phase. We will assume that the transfer takes place with the system in thermal and hydrostatic equilibrium, so that $dT = 0$ and $dP = 0$. The criterion for equilibrium in this case is

$$dG = dG^\alpha + dG^\beta \leqslant 0, \tag{5.26}$$

or since T, P, and $N_{j \neq i}$ are all constant and $dN_i^\beta = -dN_i^\alpha$,

$$-\mu_i^\alpha dN_i^\alpha + \mu_i^\beta dN_i^\alpha \leqslant 0 \tag{5.27}$$

If we assume the change to be a *natural* one, then we see that

$$\mu_i^\alpha > \mu_i^\beta, \tag{5.28}$$

or, in other words, material flows from the phase in which it has a higher chemical potential into the one in which its chemical potential is lower. Alternatively, if we suppose the system to be in equilibrium, and the process to be reversible, we have

$$\mu_i^\alpha = \mu_i^\beta \tag{5.29}$$

as the condition for distributive equilibrium. The importance of this relation cannot be overemphasized. It is the basis for phase equilibrium in all systems. In any case where two or more phases coexist at equilibrium, the chemical potential of each component of the system must be the same in all coexisting phases. Note, however, that this relation concerns the value of a *given* component in each of the phases. There is no constraint on the relative values of the chemical potentials of different components. The value of μ_i is independent of the value of μ_j insofar as the distribution of components among the phases present is concerned.

CHEMICAL EQUILIBRIUM

In order to develop a relationship *among* the chemical potentials of the various species present, we must relax the restriction that we have made up to now prohibiting chemical reactions among the components of the system. In order to treat this question of chemical equilibrium, we may consider a process, in a system at constant temperature and pressure, which may consist of several phases, in which a chemical reaction among the components can take place. For this process, we may write

$$dG = -SdT + VdP + \sum_\alpha \sum_i \mu_i^\alpha dN_i^\alpha \leqslant 0, \tag{5.30}$$

or, since $dT = 0$ and $dP = 0$ we obtain

$$dG = \sum_\alpha \sum_i \mu_i^\alpha dN_i^\alpha, \tag{5.31}$$

where the double sum is taken over all phases present and over all components in the system.

In order to account for the stoichiometry of the chemical reaction involved (that is, to account for the fact that the number of molecules of one component required for the reaction may not be the same as the number of molecules of some other component), we represent the changes taking place in the reaction as

$$\sum_\alpha \sum_i v_i^\alpha A_i^\alpha \to \sum_\beta \sum_j v_j^\beta B_j^\beta, \tag{5.32}$$

in which v_i moles of components A_i in the various phases present react to form v_j moles of components B_j in the phases present. For example, in the reaction

$$2H_{2(gas)} + O_{2(gas)} \to 2H_2O_{(liquid)}$$

We have

$$A_1^\alpha = H_2, \qquad v_1^\alpha = -2, \qquad \alpha_1 = \text{gas}$$
$$A_2^\alpha = O_2, \qquad v_2^\alpha = -1, \qquad \alpha_2 = \text{gas}$$
$$B_1^\beta = H_2O, \qquad v_1^\beta = 2, \qquad \beta_1 = \text{liquid},$$

in which the sign of v_i is taken to be negative for reactants and positive for products. We now let the reaction proceed by some small increment, $d\xi$. In this process, dN_i of each species reacts, where

$$dN_i = v_i d\xi. \tag{5.33}$$

Thus

$$dG = \sum_\alpha \sum_i \mu_i^\alpha dN_i^\alpha \leqslant 0 \tag{5.34}$$

$$= \sum_\alpha \sum_i \mu_i^\alpha v_i^\alpha d\xi + \sum_\beta \sum_j \mu_j^\beta v_j^\beta d\xi, \tag{5.35}$$

or

$$d\xi \left(\sum_\alpha \sum_i \mu_i^\alpha v_i^\alpha + \sum_\beta \sum_j \mu_j^\beta v_j^\beta \right) \leqslant 0. \tag{5.36}$$

Since $d\xi$ is arbitrary, if the system was not in chemical equilibrium before the process took place, we have (since the v_i^α are by our convention negative and the v_j^β positive)

$$\sum_i \mu_i^\alpha |v_i^\alpha| > \sum_j \mu_j^\beta |v_j^\beta|, \tag{5.37}$$

and the reaction is spontaneous. If the system was in equilibrium prior to the process considered, we obtain

$$\sum_i \mu_i^\alpha |v_i^\alpha| = \sum_j \mu_j^\beta |v_j^\beta|, \tag{5.38}$$

and no further net reaction will take place. The above expression is thus the criterion for chemical equilibrium.

Note that in this case we have developed a relation *among* the various μ_i. That is, for the case of chemical equilibrium we have

$$\mu_i = f(\mu_{j \ne i}). \tag{5.39}$$

This is in contrast to the previous case of distributive equilibrium in which the chemical potential of a given species was the same in all phases present, but independent of the μ_j of all other species present.

GENERAL CRITERIA FOR EQUILIBRIUM

Finally, let us look at an example of a multiphase system, in which all of the criteria for equilibrium developed above are met. We will consider a system composed of H_2, O_2, and H_2O molecules at a total pressure of one atmosphere and a temperature of 300 K. We will assume that the system is in overall equilibrium and that the relative amounts of the various components in the system are such that two phases are present. These are an α phase consisting of H_2, O_2, and H_2O molecules in a gas phase and a β phase consisting of liquid H_2O with dissolved O_2 and H_2 molecules. If the system is in *thermal* equilibrium, then

$$T^\alpha = T^\beta = T. \tag{5.40}$$

If the system is in *hydrostatic* equilibrium, then

$$P^\alpha = P^\beta = P. \tag{5.41}$$

If the system is in *distributive* equilibrium, then

$$\mu_{H_2}^\alpha = \mu_{H_2}^\beta, \quad \mu_{O_2}^\alpha = \mu_{O_2}^\beta, \quad \mu_{H_2O}^\alpha = \mu_{H_2O}^\beta. \quad (5.42)$$

Finally, if the system is in *chemical* equilibrium, then

$$\mu_{H_2O}\nu_{H_2O} = \mu_{O_2}\nu_{O_2} + \mu_{H_2}\nu_{H_2}. \quad (5.43)$$

THE GIBBS PHASE RULE

Before we leave the subject of equilibrium in multiphase systems, we will develop one more relationship of great significance, namely, the *Gibbs Phase Rule*.

To do this, we return to the case of a single-phase system in internal equilibrium and recall that for this case we could write the following Gibbs–Duhem equation relating the $C + 2$ intensive variables of the system:

$$SdT - VdP + \sum_i N_i d\mu_i = 0. \quad (5.44)$$

Since the $C + 2$ variables are related by this equation, only $C + 1$ of them are independent and, from our definition of degree of freedom, constitute degrees of freedom.

If we now generalize to the case of a system composed of several phases, *each* of which is in internal equilibrium and *all* of which are in mutual distributive equilibrium, we can write a Gibbs–Duhem equation for each phase:

$$S^\alpha dT^\alpha - V^\alpha dP^\alpha + \sum_i N_i^\alpha d\mu_i^\alpha = 0, \quad (5.45)$$

$$S^\beta dT^\beta - V^\beta dP^\beta + \sum_i N_i^\beta d\mu_i^\beta = 0, \quad (5.46)$$

and so on. If the system is in overall equilibrium, then

$$T^\alpha = T^\beta = \cdots = T, \quad (5.47)$$

$$P^\alpha = P^\beta = \cdots = P, \quad (5.48)$$

$$\mu_1^\alpha = \mu_1^\beta = \cdots = \mu_1, \quad (5.49)$$

$$\mu_2^\alpha = \mu_2^\beta = \cdots = \mu_2, \quad (5.50)$$

and so on.

Thus, although we have increased the number of phases present, and consequently the number of Gibbs–Duhem equations that the system must

48 GENERAL THEORY OF PHASE STABILITY

satisfy, we have not increased the number of intensive variables that apply to the system, because of the constraints on the intensive variables associated with the assumption of equilibrium among the phases. Consequently, since we still have $C + 2$ intensive variables and we have P Gibbs–Duhem equations, where P is the number of phases present, the number of degrees of freedom possessed by the system is

$$DOF = C + 2 - P. \tag{5.51}$$

This is the *Gibbs Phase Rule*. It describes the relationship between the number of components in the system, the number of phases present at equilibrium, and the number of intensive variables that may be changed without changing the number of phases present. If we realize that the number of degrees of freedom can never be less than zero, we see that this rule limits the number of possible coexisting phases in a system to two more than the number of components. For example, a one-component system in equilibrium will never exist as more than three phases, and it will exist as three phases only at well-defined values of pressure and temperature. We will see applications of the Phase Rule when we consider phase equilibrium in detail later.

BIBLIOGRAPHY

R. T. DeHoff, *Thermodynamics in Materials Science*, McGraw-Hill, New York, 1993, Chapter 5.

E. A. Guggenheim, *Thermodynamics*, North Holland, 1957, Chapters 1 and 2.

C. C. Kittel and H. Kroemer, *Thermal Physics*, 2nd ed., W. H. Freeman & Co., San Francisco, 1980, Chapter 9.

D. V. Ragone, *Thermodynamics of Materials*, Volume 1, John Wiley & Sons, New York, 1995, Chapter 5.

H. Reiss, *Methods of Thermodynamics*, Blaisdell, New York, 1965, Chapter 8.

PROBLEMS

5.1 Give examples of systems that are:
 (a) In hydrostatic equilibrium, but not in distributive equilibrium.
 (b) In distributive equilibrium, but not in chemical equilibrium.
 (c) In distributive equilibrium, but not in thermal equilibrium.

5.2 We will see in Chapter 24 that the equilibrium composition for many chemically reacting systems depends on the total pressure in the system.
 (a) Why is this true in general?
 (b) Is this the case for the reaction

$$CO + H_2O \rightleftharpoons H_2 + CO_2?$$

5.3 The Gibbs Phase Rule states that $DOF = C + 2 - P$ and that consequently the maximum number of phases that can coexist at equilibrium is $C + 2$.

(a) Give examples of equilibria involving the maximum possible number of phases for one- and two-component systems.

(b) Why is the maximum possible number of coexisting phases not seen on typical two-component phase diagrams?

PART II

FUNDAMENTALS OF STATISTICAL THERMODYNAMICS

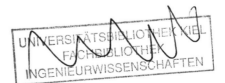

CHAPTER 6

BASIS OF STATISTICAL THERMODYNAMICS

Let us begin our discussion of statistical thermodynamics by considering the questions of what statistical thermodynamics is and why we need it.

WHAT IS STATISTICAL THERMODYNAMICS?

Up to this point, we have developed the laws of classical thermodynamics and have applied them to systems in general to determine criteria for equilibrium. The description of the state of the system has been in terms of macroscopic thermodynamic state functions. While the relations deduced are completely general and apply to any system consisting of a large number of particles, of order 10^{20}, they say nothing about the behavior of the system on a molecular level. The properties of matter on a molecular level can, however, be described in terms of quantum mechanics. This approach can be used to describe the behavior of systems containing a small number of fundamental particles, of order 10, in terms of the microscopic energy states allowed to each particle. Statistical thermodynamics is the discipline that allows us to apply the principles of quantum mechanics to systems containing large numbers of particles, and thus it forms the bridge between the microscopic treatment of quantum mechanics and the macroscopic treatment of classical thermodynamics.

The utility of the statistical thermodynamic treatment is twofold. First, we will see that we can relate the state functions developed in our treatment of classical thermodynamics to properties and processes taking place on an atomic scale. For example, we will develop expressions for properties such as temperature, internal energy, and entropy in terms of a molecular description of the system studied. Second, we will show how these macroscopic thermodynamic properties can be obtained from a knowledge of the properties of the individual particles that make up the system. We will be able to calculate values for parameters such as the free energies and the heat capacities, and the changes in these values that accompany a thermodynamic process, without recourse to a direct experiment. For example, we know from our development of classical thermodynamics that

$$C_V = \left(\frac{\partial E}{\partial T}\right)_V. \tag{6.1}$$

However, we have no means short of a direct experiment to determine the value of C_V to be expected in a given system at a given temperature. Statistical thermodynamics will provide us with methods for calculating C_V given only a knowledge of the system behavior on the molecular level.

BASIC APPROACH

Let us begin our development of statistical thermodynamics by stating a few definitions, and in the process setting up the problem that must be solved if we are to describe the properties of a macroscopic system in terms of microscopic parameters. First let us recall the definitions of macrostate and microstate: A *macrostate* of a system is described in terms of a few macroscopic thermodynamic coordinates. The molecular nature of the system is unimportant in this description; a *microstate* is described in terms of the positions and momenta of individual particles—many such parameters must be specified in order to describe the microstate of a system.

In talking about a *macrostate* of a system, we will always be concerned with a macroscopic system—that is, a system containing a large number of particles. This is necessary for two reasons. First, in most systems of interest, because there are interactions among the molecules of the system, the intensive properties that describe the macrostate of the system are not constant as the density of the system changes. For example, for a real gas at constant volume and temperature, the pressure is not linearly related to the number of molecules in the system due to intermolecular attractions and repulsions. Second, some of the properties of a macroscopic system have no meaning when applied to a single molecule. For example, one cannot talk about the "temperature" or "entropy" of a single molecule. These functions are properties of the collection of molecules that make up a macroscopic system, and consequently they have meaning only when applied to a macroscopic system.

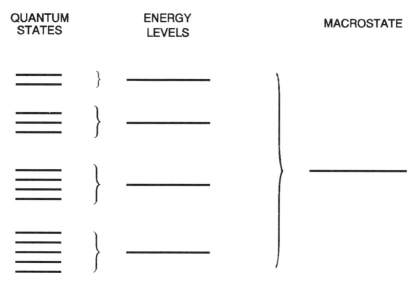

Figure 6.1 The hierarchy of quantum states, energy levels, and macrostates. In general, many quantum states (microstates) will be consistent with a given macrostate. Quantum states having the same energy, but differing in some other observable way, are grouped in energy levels.

Let us consider the question of microstates in more detail. In general, many *microstates*—or, as we will often call them, *quantum states*—of a system correspond to any single *macrostate* or thermodynamic state of a given macroscopic system. We can set up a hierarchy of states as shown in Figure 6.1. The lowest level is the individual *microstate* or *quantum state*. It is often convenient to group these *quantum states* into *energy levels*, where all of the quantum states having the same total energy are lumped together. At the top of the hierarchy is the macrostate represented by, and consistent with, all of the microstates. For future reference, we will designate the number of quantum states associated with a given energy level, say E_i, as Ω_i. That is, the energy level E_1 will have Ω_1 quantum states associated with it, the energy level E_2 will have Ω_2 states, and so on. The Ω_i of a level is often referred to as the *degeneracy* of the level E_i. It is important to keep in mind also that a macroscopic system of constant E—that is, an isolated system—has only one energy level available to it. All other systems will have many available energy levels.

Our concept of microstates also differs from that of macrostates in that we can consider microstates of individual particles or molecules as well as microstates of macroscopic systems. We can also, in some cases, assume that the microstates available to a given particle in a system are independent of the presence of the other particles in the system. We will see the consequences of this later.

RELATION OF MACROSCOPIC TO MICROSCOPIC DESCRIPTIONS

Let us now turn to the problem of how we can describe the thermodynamic state of a macroscopic system in terms of the Ω_i and E_i of the individual quantum states consistent with that thermodynamic state. The basic problem is that in any real system at equilibrium, the equilibrium is a dynamic one. Particles are moving and exchanging energy with one another and with the surroundings. Consequently the quantum state of the system changes rapidly with time. That is, over a period of time a system passes through a large number of the available quantum states. Over a large enough period, a system passes through *all* of the available quantum states. Consequently, to describe the macroscopic thermodynamic state in terms of the quantum states, we must find some way of averaging the system properties over all of the quantum states that the system has occupied over a period of time. This averaging process constitutes the central problem of statistical thermodynamics. In order to solve it, we must know *what quantum states* are available and *what is the probability* of occupancy of each state as a function of time.

ENSEMBLES

The method that we will use to obtain the properties of the macrostate of a system from information about the microstates consistent with that macrostate is known as the *ensemble method*, first put forth by Gibbs.

Before we go into the development of this method, let us first show why we need such a mechanism in order to solve the problem at hand. What we are going to deduce from the statistical mechanical approach is a series of expressions for the time-average values of the *mechanical variables* of the macroscopic system under consideration. By *mechanical variables*, we mean those variables such as P, E, V, and N, where we can in principle, and often in practice, think of each particle in the system as making some unit contribution to the value of that property in the macroscopic system. These variables may be contrasted to the *nonmechanical variables*, such as T, S, F, G, and μ, which are properties of the system as a whole and cannot be thought of in terms of a contribution per particle. Statistical mechanics cannot give us information on these nonmechanical variables directly. We must have recourse to classical thermodynamics to develop the connection between the properties of the microstates and these *nonmechanical variables*.

Let us look now at the problem of determining the long-time average of a mechanical variable—for example, pressure in a gas. The pressure in a gaseous system arises from impacts of the molecules of the system with the system walls. Since these impacts are random, the instantaneous value of the pressure will fluctuate with time, the relative magnitude of the fluctuation being inversely proportional to the number of molecules in the system. Ideally, to calculate the long-term average pressure in the system, we should measure the instantaneous

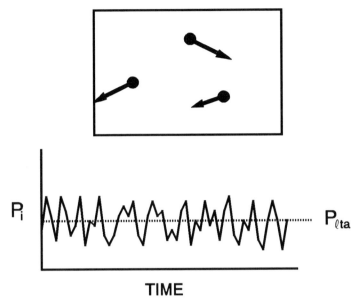

Figure 6.2 Schematic representation of the instantaneous pressure, P_i, as a function of time for a system containing a small number of gas molecules. Fluctuations are a result of the random nature of the momentum transfer between the molecules and the system walls. Pressures fluctuate about a long time average pressure, P_{lta}.

value of the force per unit area on the system walls over a long period of time, and then average these readings to obtain the desired long-time average value of the pressure. As an example of such a process, consider the system shown in Figure 6.2. Here we have a system which is a gas contained in a box of fixed volume. For simplicity we assume that the system contains only three molecules. In this case it is clear that there will be significant fluctuations in the instantaneous pressure, because there will be times when all three particles are in contact with the walls, and times when none of them will be. A plot of the instantaneous pressure versus time would thus look like the one shown in the figure.

In practice, the problem that arises in an attempt to construct a graph such as that shown in Figure 6.2 is that, in a macroscopic system, which contains on the order of 10^{20} particles, there is no computational technique which is capable of specifying the expected behavior of the instantaneous pressure with time or of carrying out the required averaging process.

To get around this problem, we develop a subterfuge known as an *ensemble*. With the help of postulates connecting the long-time average of a mechanical variable in the real system to the ensemble average of that same property, we can then use the ensemble to deduce the desired properties of the real system.

At this point it may be useful to point out the difference in logic between the classical thermodynamic approach to the behavior of systems and the statistical mechanical approach. In classical thermodynamics, the first step was to perform a series of experiments on systems under various constraints. One then observed the results of these experiments, and on the basis of these observations developed a small number of very general empirical laws, the laws of classical thermodynamics. In the statistical mechanical approach, however, one begins by making a model of the system or process of interest. The expected behavior of the model is then stated in terms of postulates, which are basically *ad hoc* assumptions. Calculations and experiments are then carried out to determine whether the application of the postulates to the model produces results that are in accord with experimental observation. In the long run, the justification for the use of the postulates is that they provide an acceptably accurate description of the behavior of real systems. This is essentially the same justification that one has for the laws of classical thermodynamics.

Consider next the question of just what an ensemble is, and what the use of this concept permits us to do that cannot be done otherwise. An *ensemble* is a mental collection of a large number of macroscopic systems, each of which is a duplicate, on a macroscopic scale, of the system whose properties we are trying to deduce. Each of the members of the ensemble is exposed to the same external conditions as the real system on which it is based, and the whole collection of ensemble members, plus any reservoir that is required to maintain the external conditions, is considered to be isolated from the rest of the universe.

In this context, the term "large number" means a number large compared to the number of microstates consistent with the given macrostate of the real macroscopic system under consideration. This definition of "large number" is required to ensure that, at any instant, each of the available microstates, or quantum states, will be observed in an appropriate number of members of the ensemble.

Let us look at the concept of the ensemble in more detail. Consider first the properties of the various members of the ensemble. Although all of these members are identical on a macroscopic scale, at any instant different ensemble members can and will be in different microstates consistent with the given macrostate. For example, if members of the ensemble can interchange energy with one another, but not mass, then each member has fixed N, V, and T. Other properties—for example, energy or pressure—will differ for different microstates of the real system, and thus will differ among the various members of the ensemble.

Since the properties not held constant in the description of the system will differ from member to member in the ensemble, we can, at least in principle, talk about an average of each of these properties taken over all the members of the ensemble, or an *ensemble average value* of properties like energy or pressure in a system of constant N, V, and T. For instance, let us return to the previous example of a system of constant N, V, and T containing only three particles. We showed at that time that the instantaneous value of the pressure

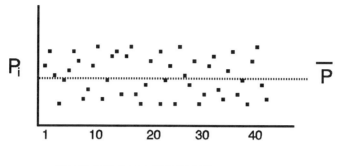

ENSEMBLE MEMBER NUMBER

Figure 6.3 The instantaneous pressures in the various systems of an ensemble of members in which pressure is not fixed in the description of the system. According to the second postulate, the ensemble average pressure, P, is equivalent to P_{lta} in Figure 6.2.

in the real system would fluctuate with time. We could also consider an ensemble made up of many members, each of which had the same N, V, and T as the system previously discussed. If we were to look at this ensemble at any instant, we would find that the members of the ensemble showed a range of instantaneous pressures. A plot of P_i, the instantaneous pressure in a given ensemble member, against some index that enumerated the members of the ensemble, would yield a plot similar to that shown in Figure 6.3. Note that this plot shows a range of values of P_i, as did the plot of P_i versus time that we used to discuss the fluctuations of P_i for the real system. We will return to this point shortly.

TYPES OF ENSEMBLES

Before we return to the previous point, however, let us digress slightly to talk about the various kinds of ensembles we will deal with, and to give rough physical examples of what an ensemble would look like. Consider first an ensemble based on a prototype that has a fixed N, V, and E—that is, an isolated system surrounded by rigid, thermally insulating walls. Such an ensemble is called a *microcanonical ensemble*. A physical example of such an ensemble would be a large number of identical-size blocks of dry ice, each in an insulated container and all contained in an insulated truck. A second type of ensemble can be formed based on a system of constant N, V, and T—that is, a closed isothermal system. Such an ensemble is called a *canonical ensemble*. A physical example of such an ensemble would be a large collection of metal cylinders of argon gas, stored in a room kept at constant temperature. Finally, we could consider an ensemble based on a system having constant μ, V, and T—that is, an open isothermal system. Such an ensemble is called a *grand*

60 BASIS OF STATISTICAL THERMODYNAMICS

canonical ensemble. A physical example in this case might be a large number of identical-size sugar cubes, held in a container of a saturated solution of sugar in water at a constant temperature.

POSTULATES

Now that we have developed the concept of an ensemble and have defined the three most common types of ensemble in terms of the macroscopic variables necessary to describe them, let us turn to the question of using these ensembles to develop the statistical mechanical expressions that describe the properties of the real prototype system in terms of properties of the appropriate ensemble. We begin by stating two postulates concerning ensembles:

1. For an isolated system, having a constant N, V, and E, all accessible microstates (quantum states) of the system, which necessarily have the same energy, are equally probable. Alternatively, the instantaneous state of an isolated system is equally likely to be any of the accessible states.
2. For any ensemble, the long-time average of any mechanical property of the real system on which the ensemble is based is equal to the properly weighted mean of that property taken over the appropriate ensemble.

Note that the word "accessible" in the first postulate means "quantum mechanically allowed." Recall that the mechanical properties referred to in the second postulate are those properties that can be thought of in terms of a unit contribution from each particle in the system — that is, variables like P, E, and N.

The consequences of these two postulates, taken together, for the case of a microcanonical ensemble based on a system of constant N, V, and E are as follows:

1. If a system of the ensemble is selected at random, the probability that it will be in a given quantum state is the same for all accessible quantum states of the original system.
2. Over a long period of time, the original system spends equal amounts of time in each of the quantum states available to it.
3. Each accessible quantum state of the real system is represented by the same number of members of the ensemble.

In order to see the significance of the two postulates stated, in terms of the behavior of an ensemble, let us look at a numerical example. Let us assume a system containing two physically distinguishable particles, a and b, and assume that each particle must have one of three possible combinations of position and momentum. That is, particle a may have position–momentum coordinates 1, 2, or 3, and particle b may have position–momentum coordinates 1, 2, or 3.

Assume for the moment that the system is a closed isothermal system, in which energy can be interchanged with the surroundings, but not mass. If we designate the allowed quantum states of this system by the position–momentum coordinates of particles a and b, respectively, the available quantum states of the system are (1, 1), (1, 2), (2, 1), (2, 2), (3, 1), (1, 3), (2, 3), (3, 2), and (3, 3). If we assume that the energies associated with position–momentum coordinates 1, 2, and 3 are all different and increase in the order 1, 2, and 3, then the system quantum states listed above will have different energies. We can list the quantum states of the system, grouping those of equal energy, as

$$(1, 1)$$
$$(1, 2) \ (2, 1)$$
$$(2, 2)$$
$$(1, 3) \ (3, 1)$$
$$(2, 3) \ (3, 2)$$
$$(3, 3).$$

Note that each combination of two numbers represents an allowed microstate (or *quantum state*, or *energy state*) of the real system. Each combination with the same total energy—for example, (2, 1), (1, 2)—is in a given *energy level*. The number of states in a given level is called the *degeneracy* of that level, $\Omega_i(N, V, E)$. That is, state (1, 1) is in the first energy level; the degeneracy of that level, Ω_1, equals 1. States (1, 2) and (2, 1) are in the second energy level; the degeneracy of that level, Ω_2, equals 2, and so forth.

Now let us form an ensemble representing this system, as shown in Figure 6.4a. Here we have enumerated the members of the ensemble by giving each member a letter designation A, B, C, and so on. We identify the quantum state of each ensemble member at any instant by the two-number designation defined above. If we look at this ensemble at any instant in time, we will see that *each* of the ensemble members is in *one* of the allowed quantum states of the real system. We will also see that, since we are looking at an ensemble representing a closed isothermal system, different ensemble members will have different energies at any instant. The second postulate stated above says that if we take a properly weighted average of these energies over all the members of the ensemble, the resulting *ensemble average energy*, $\bar{E}$, will be equal to the long-time average value of the energy, E, of the real system on which the ensemble is based. Note that if we look at this same ensemble at some other time, as shown in Figure 6.4b, the various ensemble members will be in different states than they were the first time we looked. However, the *average* of the energy, or any other mechanical variable, over the members of the ensemble will not have changed. We can define this ensemble average in terms of the probability, p_i, that a given ensemble member is in a given quantum

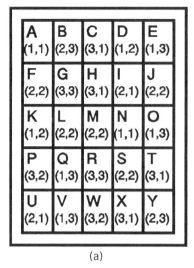 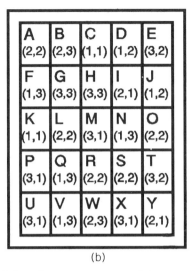

Figure 6.4 Schematic representation of an ensemble based on a system containing two distinguishable particles, each of which may have energy values 1, 2, or 3, with the systems in thermal contact with one another. (a) Distribution of the systems of the ensemble among the allowed system microstates at some time, t. (b) Distribution of the systems of the same ensemble among the allowed microstates at some other time, t'. The states of the individual systems have changed with time, but the distribution over the allowed states has not.

state. For the case of the ensemble average energy, for example, we would have

$$\bar{E} = \sum_i p_i E_i, \tag{6.2}$$

where the E_i are the energy values appropriate to each of the quantum states of the real system, and the sum is taken over all available quantum states of the real system.

Alternatively, we can look at the same real system, but this time isolated from its surroundings. In this case the system will have a constant N, V, and E, with the value of E being whatever value of E the system had at the instant it was isolated from its surroundings. In this case the only allowed quantum states of the system will be those consistent with the given value of E, and a microcanonical ensemble formed on the basis of this isolated system will have its members in these available quantum states, all having the same energy. For example, let us suppose that when the system was isolated, it had a total energy appropriate to the fourth energy level of the system. The only quantum states consistent with this total energy are states $(3, 1)$ and $(1, 3)$. Thus if we look at the microcanonical ensemble based on this isolated system, at any instant, all of the ensemble members will be in either state $(1, 3)$ or state $(3, 1)$. Moreover, according to the first postulate, the number of members in state (1, 3) will be equal to the number of members in state (3, 1).

We could equally well relax the constraint on the number of particles in the system and consider a grand canonical ensemble based on this system. In this case the ensemble would include, in addition to the two-particle states already listed, states with more or fewer particles, such as (1), (1, 2, 3), or (1, 2, 2, 3). According to the postulates previously stated, all of these states, appropriately weighted, would have to be considered in determining the ensemble average properties of this ensemble.

BIBLIOGRAPHY

D. L. Goodstein, *States of Matter*, Prentice-Hall, Englewood Cliffs, NJ, 1975, Chapter 1.
T. L. Hill, *Introduction to Statistical Thermodynamics*, Addison-Wesley, 1960, Chapter 1.
C. C. Kittel and H. Kroemer, *Thermal Physics*, 2nd ed., W. H. Freeman & Co., San Francisco, 1980, Chapters 1 and 2.
E. L. Knuth, *Introduction to Statistical Thermodynamics*, McGraw-Hill, New York, 1966, Chapter 6, Appendix D.
W. G. V. Rosser, *An Introduction to Statistical Physics*, Ellis Horwood, Chichester, UK, 1982, Chapter 2.

PROBLEMS

6.1 Consider a system composed of four physically distinguishable particles, each of which may have one of three possible position–momentum coordinates, with $E_1 < E_2 < E_3$.

(a) List the available system quantum states for this system if it is in contact with a heat reservoir.

(b) Group the available states according to energy levels.

(c) If the system is isolated at a total energy of $E_1 + 2E_2 + E_3$, what are the available energy states?

(d) If the system is modified so that particles can flow into and out of the system, how does this change the number of available energy states?

6.2 We have considered that in order to measure the long-time average value of a macroscopic property of a system, we should average over all possible quantum states of the system. If a typical system has available on the order of e^N quantum states, how long would it take for the system to pass through all of these states, assuming that it passes through each state once and only once; that the rate of passage is on the order of $10^{13} \sec^{-1}$ (typical of the vibrational frequency in a solid); and that N is on the order of Avogardo's number? Compare your answer with the age of the universe, approximately 10^{10} years.

CHAPTER 7

EVALUATION OF PROBABILITIES

We are now in a position to use the two postulates stated previously to derive expressions for the properties of various ensembles. We will do this in a completely general way, so that the results obtained will be applicable to any system. In subsequent chapters we will apply these general results to specific systems.

APPLICATION OF POSTULATES

What we wish to determine is the time-average values of the mechanical variables appropriate to a given real macroscopic thermodynamic system. The second postulate states that the desired long-time average value of any property in the real system is equivalent to the ensemble average of the same property; that is,

$$P \leftrightarrow \bar{P}, \quad E \leftrightarrow \bar{E}, \quad N \leftrightarrow \bar{N}, \tag{7.1}$$

where we will use the symbol $\leftrightarrow$ to indicate "is equivalent to." We are now faced with the problem of how to evaluate the desired ensemble average properties. To do this, we must determine (a) the value of the desired property associated with each available quantum state of the real system and (b) the probability that a given system of the ensemble is in each of the available quantum states or, alternatively, the fraction of the ensemble members in each quantum state at any instant. Stated mathematically, we must determine, for example:

$$E \leftrightarrow \bar{E} = \sum_i p_i E_i, \tag{7.2}$$

where the p_i are the probabilities of occurrence of each of the i available quantum states and the E_i are the energies of these states.

This problem may be considered as two separate problems, namely, calculation of the E_i and calculation of the p_i. For any given system, the E_i will depend on the particular physical nature of the system. In order to calculate the E_i we must have recourse to quantum mechanics. The p_i, on the other hand, can be treated generally. We can develop expressions for the p_i to be expected in various systems using an approach that does not involve any assumptions about the exact physical nature of the system being considered. It is this problem that we will attack next.

THE MICROCANONICAL ENSEMBLE

We will solve this problem of finding a general expression for the p_i in the three ensembles that we have discussed previously. Let us look first at the case of the microcanonical ensemble representative of a system of fixed N, V, and E. In this case, the determination of the p_i is trivial. We have already taken as a postulate that all states of a given system that have the same N, V, and E are equally probable. Since, in any isolated system, there will be some fixed number, $\Omega(N, V, E)$, of quantum states available, the probability of any given state is simply

$$p_i = \frac{1}{\Omega}. \tag{7.3}$$

The expressions for the ensemble average properties, such as pressure or energy, are simply

$$\bar{P} = \frac{1}{\Omega} \sum_i P_i \leftrightarrow P, \qquad \bar{E} = \frac{1}{\Omega} \sum_i E_i \leftrightarrow E. \tag{7.4}$$

THE CANONICAL ENSEMBLE

The result above was obtained easily, but by itself it is not very useful. In order to get at concepts like temperature or chemical potential, which arise from consideration of the interaction of a system with its surroundings, we must consider the properties of ensembles based on systems that can interact with the environment or with one another. We can begin to get at this problem by developing an expression for the p_i in an ensemble representative of a closed isothermal system of fixed N and V in contact with a heat reservoir at temperature T—that is, a canonical ensemble. Note that in this case, since the real system and each member of the ensemble are in contact with a heat reservoir, energy can flow into or out of the system and the ensemble members to maintain thermal equilibrium. Consequently, the energy of the real system will fluctuate with time about some long-time average value, and the members of the ensemble, at any given instant, will be found in quantum states having different energies.

The question now is, How do we carry out the averaging in this case, or, in other words, what are the appropriate p_i for states of different E_i? In order to get at this question, we will make an additional assumption, namely:

For the states of a system of given N and V in contact with a heat reservoir at temperature T, the probability of occurrence of a given quantum state is a function of the energy of that state only. Or, stated mathematically:

$$p_i = f(E_i). \tag{7.5}$$

We note in passing that this is not really an assumption, and could be shown rigorously to follow from our previous assumptions. We will not go through this proof in detail. The logic of the assumption made is essentially that since the first postulate indicates that states of equal N, V, and E all have the same probability, the probability must be independent of any other properties of the system. Thus if the p_i depend *only* on N, V, and E while N and V are held constant, the p_i should depend only on E_i, as assumed above.

Let us apply this assumption to the members of a canonical ensemble, all of which have the same N and V but may have different E_i. Let us consider any two members of this ensemble. The probability that a given ensemble member is in a particular quantum state, say state j having energy E_j, is by our assumption

$$p_j = f(E_j). \tag{7.6}$$

The probability that some other member is in some other particular quantum state, say state k of energy E_k, is similarly

$$p_k = f(E_k). \tag{7.7}$$

Moreover, the same assumption that says that the probability that a given ensemble member is in a given quantum state is a function of the energy of that state only also allows us to state that the compound probability that any chosen ensemble member is in state j and that simultaneously another chosen

68   EVALUATION OF PROBABILITIES

member is in state k is given by

$$p_{j,k} = f(E_j + E_k). \tag{7.8}$$

We could develop this assumption rigorously, using an argument that involves forming a canonical ensemble of combined systems, each combined system consisting of two of the original systems in thermal contact with each other. A member of an ensemble representing such a combined system would have an energy that was the sum of the energies of the two parts. Since *each part* of the combined system has the same N and V as the original system, it will have the same allowed values of E_i. Thus the energy of the combined system, with one part in state j and the other part in state k, would be $(E_j + E_k)$. Since we have postulated that the probability of occurrence of a given quantum state is a function of its energy only, this leads, for the combined system, to equation (7.8).

We may also state, returning to our original canonical ensemble of members having N, V, and T, that since the ensemble contains very many members, the probability that one member is in state j is independent of the probability that some other member is in state k. If this is so, then the law of compound probability states that

$$p_{j,k} = p_j \cdot p_k, \tag{7.9}$$

or

$$f(E_j + E_k) = f(E_j) \cdot f(E_k), \tag{7.10}$$

where f is the same function for all three variables.

We are now left with the purely mathematical question of what form of $f(E_i)$ satisfies the equation above. To answer this question, let us introduce, purely for convenience in writing out the argument to follow, the dummy variables x and y such that

$$E_j \equiv x, \quad E_k \equiv y. \tag{7.11}$$

Thus,

$$f(x + y) = f(x) \cdot f(y). \tag{7.12}$$

If we take the partial derivative of this expression with respect to y, we have

$$\frac{\partial f(x + y)}{\partial y} = f(x) \cdot \frac{\partial f(y)}{\partial y}, \tag{7.13}$$

or

$$\frac{\partial f(x + y)}{\partial (x + y)} \cdot \frac{\partial (x + y)}{\partial y} = f(x) \cdot \frac{\partial f(y)}{\partial y}, \tag{7.14}$$

or

$$\frac{df(x+y)}{d(x+y)} = f(x) \cdot \frac{df(y)}{dy}. \tag{7.15}$$

(because $f(x)$ is independent of y and $f(y)$ is independent of x).

Similarly, if we differentiate the original expression with respect to x we arrive at

$$\frac{df(x+y)}{d(x+y)} = f(y) \cdot \frac{df(x)}{dx}. \tag{7.16}$$

Since the left-hand sides of these two equations are identical, we may equate the right-hand sides to yield

$$f(x) \cdot \frac{df(y)}{dy} = f(y) \cdot \frac{df(x)}{dx}, \tag{7.17}$$

or

$$\frac{1}{f(y)} \cdot \frac{df(y)}{dy} = \frac{1}{f(x)} \cdot \frac{df(x)}{dx}. \tag{7.18}$$

Since, as we have said, x and y are independent of one another, the left-hand side of this last equation is not a function of x and the right-hand side is not a function of y. The only way that the equation can be valid in general is if both sides of the equation are equal to a constant — the same constant for both sides. Thus

$$\frac{1}{f(y)} \cdot \frac{df(y)}{dy} = \frac{1}{f(x)} \cdot \frac{df(x)}{dx} = -\beta, \tag{7.19}$$

where $-\beta$ is a constant whose physical significance will be determined later. At this point we will simply note that the result of our statistical treatment is that systems of a canonical ensemble — which, by the way the ensemble is defined, are in the thermal equilibrium — have a common property that we have called $-\beta$. This may be compared to the statement of the zeroth law of classical thermodynamics which states that systems in thermal equilibrium have the same temperature. This suggests a close relation between $-\beta$ and temperature. The exact relation will be developed later.

For the moment, we will return to equation (7.19) above, which may be integrated to yield

$$f(x) = Ae^{-\beta x}, \qquad f(y) = Ae^{-\beta y}, \tag{7.20}$$

where A is an integration constant. Thus, since

$$x = E_j, \quad y = E_k, \tag{7.21}$$

we have

$$p_j = f(E_j) = Ae^{-\beta E_j}, \quad p_k = f(E_k) = Ae^{-\beta E_k}, \tag{7.22}$$

and similarly for any other allowed quantum state of the system.

Because it will be more convenient later, we will make the substitution

$$A \equiv \frac{1}{Q}. \tag{7.23}$$

Thus, in general,

$$p_i = \frac{1}{Q} e^{-\beta E_i}. \tag{7.24}$$

Finally, since

$$\sum_i p_i = 1, \tag{7.25}$$

we have

$$\sum_i p_i = \frac{1}{Q} \sum_i e^{-\beta E_i} = 1, \tag{7.26}$$

or

$$Q = \sum_i e^{-\beta E_i}, \tag{7.27}$$

where $Q = Q(N, V, \beta)$, since the E_i are a function of N and V only. $Q(N, V, \beta)$ is called the *canonical ensemble partition function*, and it represents the sum of the term $e^{-\beta E_i}$ over all of the available quantum states of the system.

THE GRAND CANONICAL ENSEMBLE

Consider next the problem of determining the probability of occurrence of the various available quantum states in an open isothermal system — that is, a system of fixed volume, V, with walls that are both thermally conducting and permeable to molecules, which is immersed in a large reservoir having a constant temperature and constant chemical potentials of all species present.

In other words, this is a system of constant μ, V, and T. If we form an ensemble based on such a system, we have what is known as a *grand canonical ensemble*.

We must now develop an expression for the fraction of the members of this ensemble in each of the available quantum states. This case is more complicated than the previous case of the canonical ensemble, because we must consider quantum states associated with different values of N as well as different values of E_i. However, if we look at the grand canonical ensemble as a collection of canonical ensembles, each with a different value of N, we can still say, based on our previous calculation, that for any particular value of N we have

$$p_i(N) = A(N)e^{-\beta E_i(N)}, \tag{7.28}$$

where $p_i(N)$ is the probability that a member of a grand canonical ensemble that contains N molecules is in the quantum state $E_i(N)$ appropriate to that value of N.

Moreover, if we make the assumption, similar to that made for the case of the energies of the allowed quantum states, that the probability that a member of the ensemble contains N molecules, irrespective of the energy of the member, is a function of N only, we may write

$$p_N = f(N), \tag{7.29}$$

where p_N is the probability that the member contains N molecules. As in the previous case it can be shown rigorously that this relation is valid. As before, we state it as an assumption merely to save time.

By reasoning similar to that applied to the p_i in the case of the canonical ensemble, we can deduce that

$$p_N = A'e^{\gamma N}, \tag{7.30}$$

or, for the case of a grand canonical ensemble based on a multicomponent system,

$$p_{N_i} = A'e^{\gamma_i N_i}, \tag{7.31}$$

where A' and the γ_i are constants, and γ_i is the same constant for component i for all systems in distributive equilibrium. (Note the similarity to the relation between β and T in the previous case; here the statistical mechanical result that systems in distributive equilibrium have the same γ_i may be compared to the classical thermodynamic result that systems in distributive equilibrium have the same chemical potentials, μ_i.)

The desired probability of occurrence of a given quantum state containing N molecules and in the energy state $E_i(N)$, which we will call $p_{i,N}$, is simply the product of the probability that the member in question contains N

molecules and that it is in the state $E_i(N)$ consistent with that N. That is,

$$p_{i,N} = p_N p_i(N), \tag{7.32}$$

or

$$p_{i,N} = A' e^{\gamma N} \cdot A(N) e^{-\beta E_i(N)}. \tag{7.33}$$

Again, since each ensemble member must be in one of the allowed quantum states, if we sum over all possible values of N and $E_i(N)$ we have

$$\sum_{i,N} p_{i,N} = 1, \tag{7.34}$$

or

$$\sum_{i,N} A' A(N) e^{\gamma N} e^{-\beta E_i(N)} = 1, \tag{7.35}$$

or, if we define

$$\Xi \equiv \frac{1}{A' A(N)}, \tag{7.36}$$

we have

$$\Xi = \sum_{i,N} e^{\gamma N} e^{-\beta E_i(N)}, \tag{7.37}$$

where Ξ is the *grand canonical ensemble partition function*.

Consequently, our final expression for the probability that a system of a grand canonical ensemble will contain N molecules, and be in the quantum state $E_i(N)$ consistent with that N, is

$$p_{i,n} = \frac{1}{\Xi} e^{\gamma N} e^{-\beta E_i(N)}. \tag{7.38}$$

Thus we have the desired expressions for the probability terms that arose from the description of systems in terms of the ensemble model. Our next task will be to determine the physical significance of the parameters β and γ and to use the resulting expressions to develop the correspondence between the classical and statistical treatments.

BIBLIOGRAPHY

L. A. Girifalco, *Statistical Physics of Materials*, John Wiley & Sons, New York, 1973, Chapter 1.

T. L. Hill, *Introduction to Statistical Thermodynamics*, Addison-Wesley, 1960, Appendix D.

C. C. Kittel and H. Kroemer, *Thermal Physics*, 2nd ed., W. H. Freeman & Co., San Francisco, 1980, Chapters 2 and 3.

E. L. Knuth, *Introduction to Statistical Thermodynamics*, McGraw-Hill, New York, 1966, Chapter 6.

J. W. Whalen, *Molecular Thermodynamics*, John Wiley & Sons, New York, 1991, Chapter 1.

PROBLEMS

7.1 A system has allowed energy states $E_1 = 1$, $E_2 = 1$, $E_3 = 2$, and $E_4 = 2$. In a canonical ensemble representing this system, we find that $\bar{E} = 5/4$. What is β?

7.2 Starting with the relations

$$\bar{E} = \sum_i p_i E_i, \qquad \bar{P} = \sum_i p_i P_i, \qquad p_i = \frac{1}{Q} e^{-\beta E_i}$$

obtain the derivatives

$$\left(\frac{\partial \bar{E}}{\partial V}\right)_{\beta,N} \quad \text{and} \quad \left(\frac{\partial \bar{P}}{\partial \beta}\right)_{V,N}$$

and show that

$$\left(\frac{\partial \bar{E}}{\partial V}\right)_{\beta,N} + \beta \left(\frac{\partial \bar{P}}{\partial \beta}\right)_{V,N} = -\bar{P}.$$

7.3 A macroscopic system of given N and V has allowed microstates $E_1 = 0$, $E_2 = 1$, $E_3 = 2$, and $E_4 = 2$. In a canonical ensemble representing this system in contact with a heat reservoir which has a temperature equivalent to $\beta = 2$, what is the probability of occurrence of the state E_3? What is the energy of the macroscopic system? How do the energies of the allowed microstates change if the volume of the system is reduced at constant N?

7.4 A system can contain 0, 1, 2, or 3 particles. In an ensemble representing this system, if $\gamma = -1$:
(a) What is the probability of observing a state with $N = 2$?
(b) What is $\bar{N}$?

7.5 A grand canonical ensemble is formed, based on a system that can contain 0, 1, 2, or 3 identical particles. Each particle can have an energy $\varepsilon = 1$ or $\varepsilon = 2$. Assume that the particle energy is independent of the presence of other particles.

74 EVALUATION OF PROBABILITIES

    (a) What are the allowed $E_i(N)$?

    (b) If $\gamma = -2$ and $\beta = 1$, what is the probability of finding a system of the ensemble in a state in which $N = 3$ and $E_i(N) = 4$?

    (c) If $\gamma = -2$ and $\beta = 1$, what is the average number of particles per system of the ensemble?

7.6 Develop an expression for the ensemble average volume, $\bar{V} = f(\Omega, V_i)$, for the systems of an ensemble based on a prototype system having constant (N, P, E). Ω is the number of available microstates and the V_i are the available system volumes consistent with the given (N, P, E).

CHAPTER 8

STATISTICAL MECHANICAL CRITERIA FOR EQUILIBRIUM

In the previous chapter, we developed expressions for the probability of occurrence of the various allowed quantum states in the members of ensembles based on various prototype thermodynamic systems. The resulting expressions contained mechanical thermodynamic variables, such as E, N, and V, and also parameters such as β and γ which we have not yet related explicitly to the functions of classical thermodynamics. We will now address the problem of determining these explicit relationships. Before considering β and γ directly, however, we will digress and develop functions based on the statistical mechanical treatment that describe the criteria for equilibrium on a statistical mechanical basis. The resulting functions will be seen to have properties similar to the classical thermodynamic properties such as S, F, and PV which we have shown previously to provide criteria for equilibrium under appropriate constraints in systems treated on a classical basis.

THE ISOLATED SYSTEM AND THE FUNCTION S'

We will begin by defining a function of the p_i previously developed for the systems of an ensemble, namely,

$$S' \equiv -k \sum_i p_i \ln p_i, \tag{8.1}$$

where k is a positive constant whose value will be determined later. Let us look at the expected behavior S' as a function of the p_i. Qualitatively, the p_i are always positive fractions. Consequently S' is always positive (or at least never negative). The lower limit on S' is the case in which only *one* quantum state is available to the system. In this case, $p_1 = 1$, $p_{i \neq 1} = 0$, and $S' = 0$. On the other hand, if very many quantum states are available, each of the p_i will be small, but since many states will be available, the term in $\ln p_i$ will be large and negative. Thus S' will be very large and positive.

Let us now look at the form of the relation for S' in general. If we differentiate the defining relation, we have

$$dS' = -k\,d\left(\sum_i p_i \ln p_i\right)$$

$$= -k\sum_i \left(p_i d\ln p_i + \ln p_i\, dp_i\right)$$

$$= -k\sum_i \left[p_i\left(\frac{1}{p_i}\right)dp_i + \ln p_i\, dp_i\right]$$

$$= -k\sum_i (\ln p_i + 1)dp_i. \tag{8.2}$$

Consider next the value of dS' associated with a change in the system described by the ensemble, in which the p_i associated with two states change slightly, in such a way that the probability of one state, say p_j, increases slightly and the probability of another state, say p_k, decreases by a corresponding amount, so as to leave the probabilities of all other states unchanged. We know that since

$$\sum_i p_i = 1, \tag{8.3}$$

we have

$$\sum_i dp_i = 0. \tag{8.4}$$

Consequently

$$dp_j = -dp_k. \tag{8.5}$$

Thus, for this case we have

$$dS' = -k(\ln p_j + 1)dp_j - k(\ln p_k + 1)dp_k$$
$$= -k(\ln p_j - \ln p_k)dp_j, \tag{8.6}$$

or

$$\frac{dS'}{dp_j} = -k(\ln p_j - \ln p_k). \tag{8.7}$$

This equation is true in general. Now let us look at the specific case of an isolated system in internal equilibrium. For this case, all of the p_i are equal.

Consequently,

$$\ln p_j = \ln p_k, \qquad (8.8)$$

and

$$\frac{dS'}{dp_j} = 0 \qquad (8.9)$$

for any small change about a point of equilibrium in an isolated system. That is, S' has an extreme value in this case. Note that we do not know yet whether this extremum is a maximum or a minimum.

We can determine which it is if we differentiate the expression for S' a second time, obtaining

$$\frac{d^2 S'}{dp_j^2} = -k\left(\frac{1}{p_j} - \frac{d \ln p_k}{dp_j}\right), \qquad (8.10)$$

or, since $dp_j = -dp_k$,

$$\frac{d^2 S'}{dp_j^2} = -k\left(\frac{1}{p_j} + \frac{1}{p_k}\right). \qquad (8.11)$$

Because k was defined as a positive constant and p_j and p_k are both positive, this second derivative is negative; thus S' is a maximum at equilibrium. Moreover, we may write for this case that

$$p_i = \frac{1}{\Omega}, \qquad (8.12)$$

giving

$$S' = -k \sum_i \frac{1}{\Omega} \ln \frac{1}{\Omega}. \qquad (8.13)$$

However, for this case we have

$$\sum_i \frac{1}{\Omega} = \sum_i p_i = 1. \qquad (8.14)$$

Consequently,

$$S' = k \ln \Omega. \qquad (8.15)$$

We thus see that in an isolated system, S' is a maximum at equilibrium and is given by the expression above. This result may be compared with the classical result, obtained earlier, that the criterion for equilibrium in an isolated system is that the entropy, S, be a maximum. This suggests a close relation between S' and S. This relation will be made explicit later.

THE CLOSED ISOTHERMAL SYSTEM AND THE FUNCTION F'

Alternatively, we can define a function

$$F' \equiv \bar{E} - \frac{S'}{k\beta}. \tag{8.16}$$

We may express this function in terms of probabilities by using the previously developed relations

$$\bar{E} = \sum_i p_i E_i \tag{8.17}$$

and

$$S' = -k \sum_i p_i \ln p_i. \tag{8.18}$$

Making these substitutions we have

$$F' = \sum_i p_i E_i + \frac{1}{\beta} \sum_i p_i \ln p_i$$

$$= \frac{1}{\beta} \sum_i p_i (\beta E_i + \ln p_i). \tag{8.19}$$

If we differentiate this expression, at constant N, V, and β we obtain

$$dF' = \frac{1}{\beta} \sum_i (\beta E_i + \ln p_i + 1) dp_i. \tag{8.20}$$

As we did before, we may consider a change in the system under consideration in this case, such that the probability of one state, p_j, increases at the expense of the probability of a second state, p_k, in such a way that the other p_i are unaffected. In this case,

$$dF' = \frac{1}{\beta}[(\beta E_j + \ln p_j + 1)dp_j + (\beta E_k + \ln p_k + 1)dp_k], \tag{8.21}$$

Again using

$$\sum_i p_i = 1, \qquad \sum_i dp_i = 0, \qquad dp_j = -dp_k,$$

we have

$$\frac{dF'}{dp_j} = \frac{1}{\beta}(\beta E_j + \ln p_j - \beta E_k - \ln p_k). \tag{8.22}$$

If we apply this result to two systems of a canonical ensemble based on a closed isothermal system, we know that

$$p_j = \frac{1}{Q} e^{-\beta E_j}, \quad p_k = \frac{1}{Q} e^{-\beta E_k}, \tag{8.23}$$

or

$$\ln p_j = -\beta E_j - \ln Q, \quad \ln p_k = -\beta E_k - \ln Q. \tag{8.24}$$

Thus

$$\frac{dF'}{dp_j} = \frac{1}{\beta}(\beta E_j - \beta E_j - \ln Q - \beta E_k + \beta E_k + \ln Q), \tag{8.25}$$

or

$$\frac{dF'}{dp_j} = 0. \tag{8.26}$$

Thus we have an extremum in F' in a closed isothermal system at equilibrium. Again, we may take a second derivative, yielding

$$\frac{d^2 F'}{dp_j^2} = \frac{1}{\beta}\left(\frac{1}{p_j} + \frac{1}{p_k}\right). \tag{8.27}$$

As will be shown later, β is a positive constant. Both p_j and p_k are positive fractions; thus the second derivative is positive, and F' is a minimum in a closed isothermal system at equilibrium. This may be compared to the classical result that the Helmholz free energy, F, is a minimum in a closed isothermal system at equilibrium. Again the implied connection between F' and F will be made explicit later.

THE OPEN ISOTHERMAL SYSTEM AND THE FUNCTION $(PV)'$

Finally, we may define yet a third function, namely,

$$(PV)' \equiv -\bar{E} + \frac{S'}{k\beta} + \frac{\gamma}{\beta}\bar{N}. \tag{8.28}$$

Here we may substitute

$$\bar{E} = \sum_{i,N} p_{i,N} E_i(N)$$

$$S' = -k \sum_{i,N} p_{i,N} \ln p_{i,N} \qquad (8.29)$$

$$\bar{N} = \sum_{i,N} p_{i,N} N.$$

These substitutions yield

$$(PV)' = -\sum_{i,N} p_{i,N} E_i(N) - \frac{1}{\beta} \sum_{i,N} p_{i,N} \ln p_{i,N} + \frac{\gamma}{\beta} \sum_{i,N} p_{i,N} N$$

$$= \frac{1}{\beta} \sum_{i,N} p_{i,N} [-\beta E_i(N) - \ln p_{i,N} + \gamma N]. \qquad (8.30)$$

If we differentiate this relation, as we have done before, in this case at constant γ, V, and β we obtain

$$d(PV)' = \frac{1}{\beta} \sum_{i,N} [-\beta E_i(N) - \ln p_{i,N} + \gamma N - 1] dp_{i,N}. \qquad (8.31)$$

Again, if we consider a change in two of the probabilities p_{j,N_j} and P_{k,N_k}, leaving all other probabilities unchanged, we have

$$d(PV)' = \frac{1}{\beta} [(-\beta E_j(N_j) - \ln p_{j,N_j} - 1 + \gamma N_j) dp_{j,N_j}$$

$$+ (-\beta E_k(N_k) - \ln p_{k,N_k} - 1 + \gamma N_k) dp_{k,N_k}]. \qquad (8.32)$$

Again, since

$$\sum_{i,N} p_{i,N} = 1, \qquad \sum_{i,N} dp_{i,N} = 0, \qquad dp_{j,N_j} + dp_{k,N_k} = 0,$$

we have

$$\frac{d(PV)'}{dp_{j,N_j}} = \frac{1}{\beta} [-\beta E_j(N_j) - \ln p_{j,N_j} + \gamma N_j + \beta E_k(N_k) + \ln p_{k,N_k} - \gamma N_k]. \qquad (8.33)$$

If we evaluate this relation for two systems of a grand canonical ensemble, in which

$$p_{i,N} = \frac{1}{\Xi} e^{\gamma N} e^{-\beta E_i(N)}, \qquad (8.34)$$

or

$$\ln p_{i,N} = -\beta E_i(N) + \gamma N - \ln \Xi, \tag{8.35}$$

we have

$$\frac{d(PV)'}{dp_{j,N_j}} = \frac{1}{\beta}[\beta E_j(N_j) - \beta E_j(N_j) - \gamma N_j + \gamma N_j + \ln \Xi$$
$$+ \beta E_k(N_k) - \beta E_k(N_k) - \gamma N_k + \gamma N_k - \ln \Xi], \tag{8.36}$$

or

$$\frac{d(PV)'}{P_{j,N_j}} = 0. \tag{8.37}$$

Thus $(PV)'$ has an extremum in an open isothermal system at equilibrium. Finally, again taking a second derivative we have

$$\frac{d^2(PV)'}{dp_{j,N_j}^2} = -\frac{1}{\beta}\left(\frac{1}{p_{j,N_j}} + \frac{1}{p_{k,N_k}}\right). \tag{8.38}$$

Again, since β is positive and the $p_{i,N}$ are positive fractions, we see that the second derivative is negative and thus $(PV)'$ is a maximum in an open isothermal system at equilibrium. Again we may compare this to the classical result that the product PV is a maximum in an open isothermal system at equilibrium. Again the exact relation of $(PV)'$ to PV will be made clear later.

BIBLIOGRAPHY

J. H. Knox, *Molecular Thermodynamics*, John Wiley & Sons, New York, 1978, Chapter 4.

W. G. V. Rosser, *An Introduction to Statistical Physics*, Ellis Horwood, Chichester, UK, 1982, Chapter 6.

PROBLEM

8.1 Determine the statistical mechanical criterion for equilibrium for a system of constant (μ, P, T).

CHAPTER 9

THE CONNECTION BETWEEN STATISTICAL THERMODYNAMICS AND CLASSICAL THERMODYNAMICS

In the statistical mechanical development we have just carried out, we have developed expressions for the ensemble average values of the mechanical variables applicable to the system on which the ensemble was based. Using the postulate previously stated, we can equate these ensemble average values with the time-average values of the same property in the real system. Thus

$$E \leftrightarrow \bar{E} = \sum_i p_i E_i = \frac{1}{Q} \sum_i E_i e^{-\beta E_i}, \qquad (9.1)$$

$$P \leftrightarrow \bar{P} = \sum_i p_i P_i = \frac{1}{Q} \sum_i P_i e^{-\beta E_i}, \qquad (9.2)$$

$$N \leftrightarrow \bar{N} = \sum_{i,N} p_{i,N} N_i = \frac{1}{\Xi} \sum_{i,N_i} N_i e^{\gamma N} e^{-\beta E_i(N)}. \qquad (9.3)$$

In order for the above expressions to be of any practical use, and in order to relate the results of the statistical treatment to the nonmechanical variables of classical thermodynamics, we must determine the equivalence between the parameters β and γ and the state functions of classical thermodynamics. We have already observed in our treatments of the canonical and grand canonical ensembles that β, like T, is a criterion for thermal equilibrium and that γ, like μ, is a criterion for distributive equilibrium. We also showed, in Chapter 8, that we could define functions of these variables that had the same significance, as criteria for equilibrium, as certain of the classical thermodynamic state functions. We will now go through mathematical arguments to make these connections explicit.

THE CANONICAL ENSEMBLE AND β

We will demonstrate the equivalences desired by developing equations, based only on the statistical treatment already carried out, and demonstrating that these equations correspond, term for term, with expressions developed on a purely classical thermodynamic basis. We begin by evaluating β. To do this we will start with the statistical mechanical expression for $\bar{E}$ of the systems of a canonical ensemble:

$$\bar{E} = \sum_i p_i E_i. \tag{9.4}$$

If we consider a change in the prototype system involving either a change in temperature or a change in volume at fixed N (that is, the system is closed), we may write in general that

$$d\bar{E} = d\left(\sum_i p_i E_i\right) = \sum_i E_i dp_i + \sum_i p_i dE_i. \tag{9.5}$$

We may substitute into this relation that since

$$p_i = \frac{1}{Q} e^{-\beta E_i}, \tag{9.6}$$

we have

$$\ln p_i = -\beta E_i - \ln Q, \tag{9.7}$$

or

$$E_i = -\frac{1}{\beta}(\ln p_i + \ln Q). \tag{9.8}$$

Moreover, since $E_i = E_i(N, V)$, we obtain

$$dE_i = \left(\frac{\partial E_i}{\partial V}\right)_N dV + \left(\frac{\partial E_i}{\partial N}\right)_V dN. \tag{9.9}$$

Since for the process under consideration $dN = 0$, we may also write

$$dE_i = \left(\frac{\partial E_i}{\partial V}\right)_N dV. \tag{9.10}$$

Making these substitutions yields

$$d\bar{E} = -\left(\frac{1}{\beta}\right)\sum_i (\ln p_i + \ln Q)dp_i + \sum_i p_i \left(\frac{\partial E_i}{\partial V}\right)_N dV, \tag{9.11}$$

or

$$d\bar{E} = -\left(\frac{1}{\beta}\right)\sum_i \ln p_i\, dp_i - \left(\frac{\ln Q}{\beta}\right)\sum_i dp_i + \sum_i p_i \left(\frac{\partial E_i}{\partial V}\right)_N dV. \tag{9.12}$$

We may further reduce this equation by noting that since $\sum_i p_i = 1$, we have $\sum_i dp_i = 0$, and the second term on the right-hand side is zero. Moreover, we can make the substitution, developed earlier [equations (8.2) to (8.7)], that

$$d\left(\sum_i p_i \ln p_i\right) = \sum_i \ln p_i\, dp_i, \tag{9.13}$$

to yield

$$d\bar{E} = -\left(\frac{1}{\beta}\right) d\left(\sum_i p_i \ln p_i\right) + \sum_i p_i \left(\frac{\partial E_i}{\partial V}\right)_N dV. \tag{9.14}$$

Finally, using the classical thermodynamic relation that

$$-P = \left(\frac{\partial E}{\partial V}\right)_N, \tag{9.15}$$

where P is pressure, we may write that, for any quantum state

$$\left(\frac{\partial E_i}{\partial V}\right)_N = -P_i. \tag{9.16}$$

We thus obtain

$$d\bar{E} = -\left(\frac{1}{\beta}\right) d\left(\sum_i p_i \ln p_i\right) - \sum_i p_i P_i\, dV, \tag{9.17}$$

or, remembering that

$$dS' = -kd\left(\sum_i p_i \ln p_i\right) \quad (9.18)$$

and that

$$\bar{P} = \sum_i p_i P_i, \quad (9.19)$$

we have

$$d\bar{E} = \left(\frac{1}{k\beta}\right)dS' - \bar{P}dV, \quad (9.20)$$

the desired equation based on the statistical mechanical treatment.

This equation may be compared to the classical thermodynamic statement of the first law, in general,

$$dE = dQ - dW, \quad (9.21)$$

or, for the particular case of a closed chemical system in internal equilibrium (the case considered in the development of the expression for $d\bar{E}$),

$$dE = TdS - PdV. \quad (9.22)$$

In comparing the statistical and classical equations, we note that we have already made the identifications

$$\bar{E} \leftrightarrow E, \quad \bar{P} \leftrightarrow P, \quad dV = dV. \quad (9.23)$$

This leads us to the conclusion that

$$\left(\frac{1}{k\beta}\right)dS' \leftrightarrow dQ = TdS, \quad \text{or} \quad dS' = k\beta dQ. \quad (9.24)$$

We know, finally, that from the way we defined S' it is a function of the state of the system. Consequently dS' is exact and thus $k\beta dQ$ is exact. We also know from classical thermodynamics that the operation required to make dQ exact is division by the absolute temperature, T. Thus

$$k\beta \leftrightarrow \frac{1}{T}, \quad \text{or} \quad \beta = \frac{1}{kT}, \quad (9.25)$$

and the desired relation between β and temperature has been obtained.

We now also have complete equivalence between dS' and dS because, since $\beta \leftrightarrow 1/kT$, we have

$$dS' \leftrightarrow dS. \tag{9.26}$$

If we integrate this relation we have

$$S' \leftrightarrow S + C, \tag{9.27}$$

where C is an unknown constant, associated with the fact, discussed previously, that we have no basis for establishing an absolute value for the entropy from the laws of classical thermodynamics. Since we have shown that the minimum value for S' is zero (in a system with only one available quantum state) we will assume that $C = 0$, and thus that

$$S' \leftrightarrow S, \tag{9.28}$$

and the equivalence of the two equations from the two treatments is demonstrated.

EVALUATION OF THE STATE FUNCTIONS IN TERMS OF Q

Now that we have evaluated β we may go back and express all of the other classical thermodynamic state functions in terms of the parameters of the statistical treatment. Recall that in our discussion of criteria for equilibrium on the statistical mechanical basis we showed that

$$F' = \left(\frac{1}{\beta}\right) \sum_i p_i (\beta E_i + \ln p_i), \tag{9.29}$$

and that

$$\ln p_i = -\beta E_i - \ln Q. \tag{9.30}$$

This leads to

$$F' = \left(\frac{1}{\beta}\right) \sum_i p_i (\beta E_i - \beta E_i - \ln Q) = -\left(\frac{1}{\beta}\right) \sum_i p_i \ln Q, \tag{9.31}$$

or, since

$$\sum_i p_i = 1,$$

we have

$$F' = -\left(\frac{1}{\beta}\right) \ln Q. \tag{9.32}$$

If we recall that

$$F' = \bar{E} - \frac{S'}{k\beta}, \tag{9.33}$$

and that, classically,

$$F = E - TS, \tag{9.34}$$

and since we have shown that

$$\bar{E} \leftrightarrow E, \quad S' \leftrightarrow S, \quad \text{and} \quad \left(\frac{1}{k\beta}\right) \leftrightarrow T, \tag{9.35}$$

we have

$$F' \leftrightarrow F. \tag{9.36}$$

Thus

$$F = -kT \ln Q, \quad \text{where} \quad Q(NVT) = \sum_i e^{-E_i(N,V)/kT}, \tag{9.37}$$

and we have, finally, an expression for one of the classical thermodynamic state functions in terms of a partition function and another classical thermodynamic state function. All that remains now, in order to be able to calculate the value of F for any system at given N, V, and T, is knowledge of the value of the constant k and of the allowed values of $E_i(N, V)$. This knowledge must come from a quantum mechanical treatment of a model of the particular system of interest.

For the present, let us continue the process of relating the classical thermodynamic state functions to the partition function Q. We know that

$$\bar{E} = \sum_i E_i p_i = \left(\frac{1}{Q}\right) \sum_i E_i e^{-\beta E_i}. \tag{9.38}$$

We also know that

$$Q = \sum_i e^{-\beta E_i}.$$

Thus

$$\left(\frac{\partial Q}{\partial \beta}\right)_{N,V} = -\sum_i E_i e^{-\beta E_i}. \tag{9.39}$$

Making this substitution we have

$$\bar{E} = -\frac{1}{Q}\left(\frac{\partial Q}{\partial \beta}\right)_{N,V} = -\left(\frac{\partial \ln Q}{\partial \beta}\right)_{N,V}, \tag{9.40}$$

or, since

$$\bar{E} \leftrightarrow E \quad \text{and} \quad \beta \leftrightarrow \frac{1}{kT},$$

we have

$$E = kT^2 \left(\frac{\partial \ln Q}{\partial T}\right)_{N,V}. \tag{9.41}$$

Moreover, since

$$S = -\frac{F}{T} + \frac{E}{T}, \tag{9.42}$$

we have

$$S = k \ln Q + kT \left(\frac{\partial \ln Q}{\partial T}\right)_{N,V}. \tag{9.43}$$

Finally, since

$$P = -\left(\frac{\partial F}{\partial V}\right)_{N,T}, \tag{9.44}$$

we have

$$P = kT \left(\frac{\partial \ln Q}{\partial V}\right)_{N,T}. \tag{9.45}$$

Similar expressions can be developed for any of the other classical thermodynamic state functions, given the relationship between F and Q and the relation of the other classical state functions to F.

THE GRAND CANONICAL ENSEMBLE AND γ

Returning now to the question of establishing the correspondence between the statistical and classical treatments, we see that we are still faced with the question of the evaluation of the constant γ. Since this parameter arose in our statistical mechanical treatment of an open system, we must consider a process taking place in an open system in order to evaluate it. We begin as before with the expression for $\bar{E}$, where for this case we have

$$\bar{E} = \sum_{i,N} p_{i,N} E_i(N). \tag{9.46}$$

Again, we differentiate to get

$$d\bar{E} = \sum_{i,N} E_i(N) dp_{i,N} + \sum_{i,N} p_{i,N} dE_i(N). \tag{9.47}$$

In this case

$$p_{i,N} = \left(\frac{1}{\Xi}\right) e^{-\beta E_i(N)} e^{\gamma N}$$

$$\ln p_{i,N} = -\beta E_i(N) + \gamma N - \ln \Xi. \tag{9.48}$$

Thus

$$E_i(N) = -\left(\frac{1}{\beta}\right)(-\gamma N + \ln p_{i,N} + \ln \Xi). \tag{9.49}$$

We may also write, in general, that

$$dE_i(N) = \left(\frac{\partial E_i(N)}{\partial V}\right)_N dV + \left(\frac{\partial E_i(N)}{\partial N}\right)_V dN, \tag{9.50}$$

or, since $E_i(N)$ for a given N is not a function of N, and the second term on the right is zero, we have

$$dE_i(N) = \left(\frac{\partial E_i(N)}{\partial V}\right)_N dV. \tag{9.51}$$

Making the substitution that

$$P_i(N) = -\left(\frac{\partial E_i(N)}{\partial V}\right)_N, \tag{9.52}$$

we have

$$dE = -\left(\frac{1}{\beta}\right) \sum_{i,N} (-\gamma N + \ln p_{i,N} + \ln \Xi) dp_{i,N} - \sum_{i,N} p_{i,N} P_i(N) dV, \quad (9.53)$$

or

$$dE = -\left(\frac{1}{\beta}\right) \sum_{i,N} \ln p_{i,N} dp_{i,N} + \left(\frac{\gamma}{\beta}\right) \sum_{i,N} N dp_{i,N}$$
$$- \left(\frac{1}{\beta}\right) \sum_{i,N} \ln \Xi \, dp_{i,N} - \sum_{i,N} p_{i,N} P_i(N) dV. \quad (9.54)$$

However, we have previously shown that

$$\sum_{i,N} \ln p_{i,N} dp_{i,N} = d\left(\sum_{i,N} p_{i,N} \ln p_{i,N}\right).$$

Consequently, we may make this substitution into the first term on the right-hand side of equation (9.54). Moreover, since

$$\bar{N} = \sum_{i,N} p_{i,N} N, \quad (9.55)$$

we may substitute into the second term on the right-hand side of equation (9.54) that

$$d\bar{N} = \sum_{i,N} N dp_{i,N}. \quad (9.56)$$

The third term on the right is zero, since

$$\sum_{i,N} dp_{i,N} = 0.$$

Finally, since

$$\bar{P} = \sum_{i,N} p_{i,N} P_i(N), \quad (9.57)$$

we may substitute this quantity into the final term of equation (9.54) to yield

$$dE = -\left(\frac{1}{\beta}\right) d\left(\sum_{i,N} p_{i,N} \ln p_{i,N}\right) + \left(\frac{\gamma}{\beta}\right) d\bar{N} - \bar{P} dV, \quad (9.58)$$

This may be compared to the classical thermodynamic equation for the energy change accompanying a process in an open chemical system at equilibrium,

$$dE = TdS + \mu dN - PdV. \tag{9.59}$$

Since we already know that

$$\bar{E} \leftrightarrow E, \quad \beta \leftrightarrow \frac{1}{kT}, \quad \bar{N} \leftrightarrow N, \quad \bar{P} \leftrightarrow P, \quad dV = dV,$$

$$\text{and} \quad dS' = -kd\left(\sum_{i,N} p_{i,N} dp_{i,N}\right) \leftrightarrow dS,$$

we must have

$$\frac{\gamma}{\beta} = \mu \quad \text{or} \quad \gamma = \frac{\mu}{kT}, \tag{9.60}$$

the desired relation between the statistical parameter γ and the classical state function μ.

We may thus write that

$$\Xi = \sum_{i,N} e^{-E_i(N)/kT} e^{N\mu/kT}. \tag{9.61}$$

Equivalently, we could write that

$$\Xi = \sum_{N}\left[e^{N\mu/kT} \sum_{i} e^{-E_i(N)/kT}\right], \tag{9.62}$$

or

$$\Xi = \sum_{N} Q e^{N\mu/kT} = \sum_{N} Q(NVT) e^{N\mu/kT}. \tag{9.63}$$

(Note that Q must stay *inside* the summation, because it is a function of N.)

Finally, we will do one more manipulation and comparison. Starting with

$$S' = -k \sum_{i,N} p_{i,N} \ln p_{i,N} \tag{9.64}$$

and

$$p_{i,N} = \frac{1}{\Xi} e^{-E_i(N)/kT} e^{N\mu/kT}$$

$$\ln p_{i,N} = -\frac{E_i(N)}{kT} + \frac{N\mu}{kT} - \ln \Xi, \tag{9.65}$$

we have

$$S' = k \sum_{i,N} p_{i,N}\left(\frac{E_i(N)}{kT}\right) - k \sum_{i,N} p_{i,N}\left(\frac{N\mu}{kT}\right) + k \sum_{i,N} p_{i,N} \ln \Xi, \quad (9.66)$$

or

$$S' = \frac{1}{T}\sum_{i,N} p_{i,N} E_i(N) - \frac{\mu}{T}\sum_{i,N} p_{i,N} N + k \ln \Xi \sum_{i,N} p_{i,N}, \quad (9.67)$$

or

$$S' = \frac{\bar{E}}{T} - \frac{\bar{N}\mu}{T} + k \ln \Xi. \quad (9.68)$$

This may be compared to the classical equation

$$S = \frac{E}{T} - \frac{N\mu}{T} + \frac{PV}{T} \quad (9.69)$$

(obtained from $N\mu = G = E - TS + PV$), to yield

$$PV \leftrightarrow kT \ln \Xi. \quad (9.70)$$

Again we see that the classical parameter that is the criterion for equilibrium in the type of system under consideration is simply related to the partition function appropriate to an ensemble based on that type of system.

EVALUATION OF THE STATE FUNCTIONS IN TERMS OF Ξ

In evaluating the other thermodynamic state functions in terms of Ξ, it is simplest to recall the Gibbs–Duhem equation, which for a one-component open system is

$$SdT - VdP + Nd\mu = 0, \quad (9.71)$$

and the fact that

$$d(PV) = PdV + VdP. \quad (9.72)$$

Thus

$$d(PV) = SdT + Nd\mu + PdV, \quad (9.73)$$

or

$$S = \left(\frac{\partial (PV)}{\partial T}\right)_{V,\mu} = kT\left(\frac{\partial \ln \Xi}{\partial T}\right)_{V,\mu} + k \ln \Xi, \tag{9.74}$$

$$N = \left(\frac{\partial (PV)}{\partial \mu}\right)_{T,V} = kT\left(\frac{\partial \ln \Xi}{\partial \mu}\right)_{T,V}, \tag{9.75}$$

$$P = \left(\frac{\partial (PV)}{\partial V}\right)_{T,\mu} = kT\left(\frac{\partial \ln \Xi}{\partial V}\right)_{T,\mu} = \frac{kT}{V} \ln \Xi. \tag{9.76}$$

Again, this gives us a set of expressions for the classical thermodynamic state functions in terms of a partition function. As in the previous case, expressions for all of the other thermodynamic functions could be developed from the relation of PV to Ξ and the relations of these other functions to PV.

FLUCTUATIONS

Before we leave the question of the equivalence of the statistical and classical treatments of thermodynamics, it is useful to consider in more detail the fluctuations that we have assumed to exist in the properties of the members of an ensemble. For example, we stated that in the member systems of a canonical ensemble, the energy per member, E_i, could assume a range of values. Similarly the number of molecules in a system of a grand canonical ensemble, N, was assumed to fluctuate from system to system in the ensemble and with time in the prototype system on which the ensemble was based.

Let us first look quantitatively at the question of what value of E_i we are likely to find at any time. That is, what do the p_i look like as a function of E_i? If we look at this question in terms of energy *levels* rather that energy states, we have

$$p_k = \Omega_k p_i, \tag{9.77}$$

where the index i refers to a state having a given energy E_i and the index k refers to the energy level having that same energy. For an isolated system, only one level will be populated. For an isothermal system many levels will be populated. In most real systems, we will find that Ω_k increases as E_k increases. Consequently the overall behavior of p_k or E_k will be as shown in Figure 9.1. The combined effect of an increase in Ω_k with E_k and the decrease of p_i with increasing E_i leads to a peak in p_k at some particular value of E_k.

We are now in a position to calculate the expected magnitude of the fluctuations we are likely to observe in a real system—that is, the width of the peak in Figure 9.1. To do this, we will consider fluctuations in the energies of the systems of a canonical ensemble about the ensemble average $\bar{E}$. We will see that for macroscopic systems the magnitude of the fluctuations is small

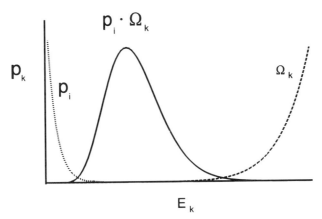

Figure 9.1 Schematic illustration of the competing effects of p_i, the probability of occurrence of a given microstate of energy E_i, and Ω_k, the number of states of energy E_i. The solid curve represents the product of $p_i \cdot \Omega_k$. In a real macroscopic system, this curve would be sharply peaked about the ensemble average energy, $\bar{E}$.

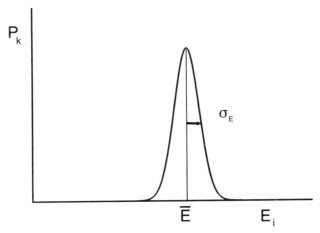

Figure 9.2 A Gaussian distribution of the probability of observing a given system total energy relative to the ensemble average energy, $\bar{E}$. σ_E is the standard deviation of E_i relative to $\bar{E}$.

compared to the mean value. We will thus assume that the probabilities of values of E_i for the systems of the ensemble follow a Gaussian distribution about $\bar{E}$, as shown in Figure 9.2. We can describe the shape of this distribution in terms of the standard deviation of E_i from $\bar{E}$, where the standard deviation is defined as

$$\sigma_E = [\overline{(E_i - \bar{E})^2}]^{1/2}, \tag{9.78}$$

or
$$\sigma_E^2 = \overline{E^2} - \bar{E}^2. \tag{9.79}$$

This σ_E^2 will always be a positive quantity.

In order to characterize the relative magnitude of an expected fluctuation in the system energy, we must determine the ratio of σ_E to $\bar{E}$ for a typical macroscopic system. We begin with the relation

$$\bar{E} = \frac{1}{Q} \sum_i E_i e^{-E_i/kT}, \tag{9.80}$$

or

$$\bar{E} \cdot Q = \sum_i E_i e^{-E_i/kT}. \tag{9.81}$$

If we differentiate this expression with respect to T at constant N and V, we have

$$\left(\frac{\partial E}{\partial T}\right)_{N,V} \cdot Q + E \left(\frac{\partial Q}{\partial T}\right)_{N,V} = \sum_i E_i \left(\frac{\partial e^{-E_i/kT}}{\partial T}\right)_{N,V}, \tag{9.82}$$

or, since

$$Q = \sum_i e^{-E_i/kT},$$

we have

$$\left(\frac{\partial Q}{\partial T}\right)_{N,V} = \sum_i e^{-E_i/kT} \frac{d}{dT}\left(\frac{-E_i}{kT}\right)$$
$$= \left(\frac{1}{kT^2}\right) \sum_i E_i e^{-E_i/kT}. \tag{9.83}$$

Thus

$$\left(\frac{\partial \bar{E}}{\partial T}\right)_{N,V} \cdot Q + \bar{E}\left(\frac{1}{kT^2}\right) \sum_i E_i e^{-E_i/kT} = \frac{1}{kT^2} \sum_i E_i^2 e^{-E_i/kT}. \tag{9.84}$$

If we now recall that

$$\left(\frac{\partial \bar{E}}{\partial T}\right)_{N,V} \leftrightarrow \left(\frac{\partial E}{\partial T}\right)_{N,V} = C_V \tag{9.85}$$

and divide through by Q, we have

$$C_V + \left(\frac{\bar{E}}{kT^2}\right)\frac{1}{Q}\sum_i E_i e^{-E_i/kT} = \left(\frac{1}{kT^2}\right)\frac{1}{Q}\sum_i E_i^2 e^{-E_i/kT}, \qquad (9.86)$$

or

$$C_V + \frac{\bar{E}^2}{kT^2} = \frac{\overline{E^2}}{kT^2}, \qquad (9.87)$$

or

$$kT^2 C_V = \overline{E^2} - \bar{E}^2. \qquad (9.88)$$

In most systems of physical interest, we find experimentally that $C_V \approx Nk$, $E \approx NkT$. Thus

$$\frac{\sigma_E}{\bar{E}} = \frac{(kT^2 C_V)^{1/2}}{\bar{E}} \approx \frac{(kT^2 Nk)^{1/2}}{NkT}$$

$$\approx \left(\frac{1}{N}\right)^{1/2}. \qquad (9.89)$$

Since in most macroscopic systems N is on the order of 10^{20}, $\sigma_E/\bar{E}$ will be on the order of 10^{-10}. That is, we are likely to observe fluctuations of the energy of a macroscopic system on the order of one part in 10^{10}. This value is clearly too small to measure by any currently available measurement technique.

Consequently, we can now state, for a macroscopic system, that even though the energy of a system of constant N, V, and T is not fixed in principle, and that all quantum mechanically allowed values of E_i are possible, as a practical matter the probability of observing a system energy measurably different from $\bar{E}$ is vanishingly small. A consequence of this is that one cannot distinguish on a physical basis a system of a canonical ensemble from a system of a microcanonical ensemble or of a grand canonical ensemble. The choice of which ensemble to use in any practical problem is thus a matter of computational convenience. Consequently, when we consider the evaluation of the properties of specific types of systems in Parts III to V of this text, we will use whichever ensemble provides the most tractable mathematical formulation, without having to worry about the environment of the system of interest. Moreover, there is no requirement that we must base our ensemble on one of the three sets of independent variables considered to this point. Any set of three independent variables can be used to describe the prototype system on which an ensemble is based. Examples of other ensembles, and the thermodynamic relations arising from them, are given in Appendix B.

BIBLIOGRAPHY

L. A. Girifalco, *Statistical Physics of Materials*, John Wiley & Sons, New York, 1973, Chapter 1.

T. L. Hill, *Introduction to Statistical Thermodynamics*, Addison-Wesley, 1960, Chapters 1 and 2.

J. H. Knox, *Molecular Thermodynamics*, John Wiley & Sons, New York, 1978, Chapter 4.

E. L. Knuth, *Introduction to Statistical Thermodynamics*, McGraw-Hill, New York, 1966, Chapter 5.

W G. V. Rosser, *An Introduction to Statistical Physics*, Ellis Horwood, Chichester, UK, 1982, Chapter 5.

PROBLEMS

9.1 Develop expressions for H, G, μ, and C_V in terms of the partition function Q and its appropriate partial derivatives.

9.2 Develop expressions for E, F, H, and G in terms of the partition function Ξ and its appropriate partial derivatives.

9.3 Verify that the equations

$$P = kT\left(\frac{\partial \ln Q}{\partial V}\right)_{T,N} \quad \text{and} \quad E = kT^2\left(\frac{\partial \ln Q}{\partial T}\right)_{V,N}$$

are equivalent to the equations

$$\bar{P} = \sum_i p_i P_i \quad \text{and} \quad \bar{E} = \sum_i p_i E_i,$$

in which

$$p_i = \frac{1}{Q} e^{-E_i/kT}, \qquad Q = \sum_i e^{-E_i/kT}.$$

9.4 Make use of the relations

$$\bar{E} = \frac{\sum_i E_i e^{-\beta E_i}}{\sum_i e^{-\beta E_i}} \quad \text{and} \quad \beta = \frac{1}{kT}$$

and the experimental fact that C_V is always positive to prove that k is a positive constant.

9.5 The following questions pertain to a system containing four particles in a box of 1-cm³ volume. The system has four allowed quantum states, $E_1 = 1$ K, $E_2 = 2$ K, $E_3 = 2$ K, and $E_4 = 3$ K, and is in contact with a heat reservoir at a temperature $T = (2/k)$ K.

(a) What kind of ensemble would one use to represent this system?

(b) What are the criteria for equilibrium in this system, on a classical basis and on a statistical mechanical basis?

(c) At any instant, what fraction of the systems of the ensemble representing this system would be found in the energy state E_3?

(d) Calculate the values of E, F, and S for this system.

9.6 Given a system of fixed N and V with allowed states $E_1 = 0$ and $E_2 = K$ energy units, develop expressions for E and C_V as a function of temperature, T. Show that as $T \to 0$, $E \to 0$ and $C_V \to 0$ and that as $T \to \infty$, $E \to K/2$ and $C_V \to 0$.

9.7 Calculate the probability of a state in an isolated system having a volume of 2 cm³ and containing 10,000 particles per cubic centimeter in which 50 particles have migrated spontaneously from one cubic centimeter of the system to the other, relative to the probability of a state in which each cubic centimeter contains 10,000 particles. (*Hint:* Start with the equation $S = k \ln \Omega$.)

CHAPTER 10

EVALUATION OF THE ALLOWED ENERGIES

Now that we have developed general expressions for the classical thermodynamic state functions in terms of the partition functions, all that remains is to develop methods for determining the appropriate values for the allowed energies of the quantum states of the system of interest. That is, now that we know, for example, that

$$F = -kT \ln Q \tag{10.1}$$

and that

$$Q = \sum_i e^{-E_i/kT}, \tag{10.2}$$

the only remaining unknowns are the E_i.

MODELS

Up to this point, the development of our statistical thermodynamic framework has been completely general. We have said, for example, that the energies of the allowed microstates, E_i, are

$$E_i = E_i(N, V), \tag{10.3}$$

but nothing more. In order to apply this general framework that we have developed to specific systems, we require values for the energies of the allowed system microstates. There is, at present, no means of obtaining the *exact* values of the allowed system energies in any real system. The best that we can do is to devise a model of the system of interest, determine the allowed energy values of this model, and by application of the model determine how closely the model describes the behavior of the real system, or class of real systems. This is the approach that we will take in this text.

In all cases, we will use models that describe the behavior of individual particles—atoms, molecules, electrons, and so on—in terms of quantum mechanics and that make assumptions concerning the way that these particles interact in systems of many particles. The basis of these processes is discussed in more detail later in this chapter.

We will see that, in general, more than one model can be used to describe the behavior of a given system. It will also become apparent that the agreement between the predictions of the model and the behavior of the real system will depend on the degree of sophistication of the model. The general observation is that the models that are the simplest mathematically provide the most limited description of system behavior, while the most mathematically sophisticated models, which by their nature are more difficult to evaluate, will provide a closer approximation to the behavior of the real system. A classic example of this balance between mathematical simplicity on the one hand and an improved description of reality on the other is provided in the two treatments of the behavior of crystalline solids presented in Chapters 14 and 15. The mathematically simple Einstein treatment reproduces the general features of the thermodynamic behavior of systems of this sort, especially insofar as the heat capacity versus temperature behavior is concerned. The more sophisticated Debye treatment provides a much closer approximation to observed behavior, but at the expense of a much more complex mathematical formulation. More exact treatments, in terms of a more detailed analysis of the allowed vibrational energy modes, are still more difficult mathematically.

The point to be made here is that, in any practical case, it will be prudent to choose the simplest model that will provide an adequate description of the system of interest. For example, the description of the ideal gas, given in Chapters 12 and 13, will be adequate to characterize the behavior of most common gases at low densities (low pressure and high temperature). Models that take into account finite molecular size and molecular interaction energies will be required only at densities high enough that these factors make a significant contribution to system properties, as is the case with nonideal gases and condensation phenomena.

QUANTUM MECHANICS

With the above discussion of models in mind, let us begin setting up the basic relations that we will need to develop and evaluate the required models. We will begin by introducing the quantum mechanical concepts needed to determine the energy states available to individual particles, and then make some approximations that permit us to determine the E_i of the macroscopic system from a knowledge of the individual particle energies.

We will not go into the development of quantum mechanics in any great detail, because that would be well beyond the scope of this text. We will, however, develop two quantum mechanical concepts that will be needed in order to determine the desired E_i, namely, the concept of quantized energy states for individual particles, and the effects that the presence of other particles have on the behavior or a given particle.

Let us look first that the question of energy states. Quantum mechanics is based, in its simplest form, on two equations: the Heisenberg uncertainty principle

$$\Delta p \times \Delta q \approx h, \quad (10.4)$$

in which p is the particle momentum, q is the particle position, and h is Planck's constant; and the deBroglie equation

$$p = mv = \frac{h}{\lambda}, \quad (10.5)$$

in which m is particle mass, v is particle velocity, and λ is the so-called deBroglie wavelength.

TRANSLATIONAL MOTION

In the simplest case, for a particle moving in field-free space, the energy of the system will be purely kinetic energy and will be given by

$$\varepsilon = \tfrac{1}{2}mv^2 = \frac{p^2}{2m} = \frac{h^2}{2m\lambda^2}. \quad (10.6)$$

As an example of this case, let us consider a particle moving in a one-dimensional box of length L, with a uniform potential inside the box. We will use this result in our treatments of the ideal monatomic gas and the electron gas. Classically, in this case, at any given time we would observe the particle at some point on the line representing the one-dimensional box, moving with some velocity v. All values of v would be allowed. According to quantum mechanics, such an observation is not possible. The consequence of the uncertainty principle is that if we have an exact specification of the particle velocity, we must be uncertain as to its position along the line, and vice versa.

104 EVALUATION OF THE ALLOWED ENERGIES

We must thus describe the particle velocity in terms of the deBroglie equation

$$v = \frac{h}{m\lambda}. \qquad (10.7)$$

That is, the particle must be thought of as having a wavelength, λ. To be an allowed state of the system, the amplitude of the wave represented by λ must be equal to zero at the boundaries of the box. (That is, the wave representing the particle must be a standing wave within the boundaries imposed on the position of the particle.) Consequently we must have

$$n\left(\frac{\lambda}{2}\right) = L, \qquad n = 1, 2, \ldots \qquad (10.8)$$

or

$$\frac{1}{\lambda} = \frac{n}{2L}. \qquad (10.9)$$

The energies allowed to the particle are thus

$$\varepsilon_n = \frac{h^2}{2m\lambda^2} = \left(\frac{h^2}{2m}\right)\left(\frac{n^2}{4L^2}\right) = \left(\frac{h^2}{8m}\right)\left(\frac{n^2}{L^2}\right), \qquad n = 1, 2, \ldots \qquad (10.10)$$

(Note that n cannot equal zero, because this would violate the uncertainty principle.) The generalization to motion in a three-dimensional box is straightforward and leads to the expression

$$\varepsilon_i = \left(\frac{h^2}{8m}\right)\left(\frac{l^2}{a^2} + \frac{m^2}{b^2} + \frac{n^2}{c^2}\right), \qquad l, m, n, = 1, 2, \ldots, \qquad (10.11)$$

where a, b, and c are the dimensions of the box.

This result has an important implication for the evaluation of the partition functions developed earlier. Since only specific values of ε_i are allowed and we have finite differences among the allowed ε_i, the number of terms that contribute to the sums that make up Ω, Q, and Ξ will be finite. This would not be the case if we treated the systems of interest classically.

ROTATIONAL MOTION

Returning to the question of expressions for allowed values of energy, we can look at the case of rotational motion. This represents, essentially, a particle translating along a circular path, and we will use this result when we consider the energy associated with the rotation of a diatomic or polyatomic molecule about its center of mass. Consideration of this problem leads to an expression for the allowed rotational energies of

$$\varepsilon_j = \left(\frac{h^2}{8\pi^2 mr^2}\right)(j)(j+1) = \left(\frac{h^2}{8\pi^2 I}\right)(j)(j+1), \qquad j = 0, 1, 2, \ldots \quad (10.12)$$

where I is the moment of inertia of the particle.

MOTION IN A POTENTIAL FIELD

When the particle of interest is in a potential field, the particle energy will include contributions that are dependent upon its position in the field. Consequently, the situation is more complicated. In this case, the total energy of the particle will be the sum of kinetic energy and potential energy terms. In order to handle this case, we must go beyond the deBroglie equation and use the Schrödinger equation:

$$\left(\frac{-h^2}{8\pi^2 m}\right)\left(\frac{\partial^2 \psi}{\partial x^2} + \frac{\partial^2 \psi}{\partial y^2} + \frac{\partial^2 \psi}{\partial z^2}\right) + U(x, y, z) = \varepsilon\psi. \quad (10.13)$$

In this equation, $U(x, y, z)$ represents the form of the potential field and ψ, the wave function, takes the place of the wavelength in the deBroglie equation. In this case, the allowed values of the wave function, ψ, the so-called "eigenvalues," are those values that give a result for ε that is not a function of position.

In order to apply the Schrödinger equation, it is necessary to assume a form for the potential and then solve the equation for the eigenvalues of ψ. For the case of harmonic motion—for example, an approximation of the vibration of an atom about its lattice point in a crystal, or the vibrational motion of the atoms of a diatomic or polyatomic molecule—we may write

$$U = fr^2, \quad (10.14)$$

where f is related to the bond strength in the crystal or the interatomic bond, and r is the displacement from the equilibrium position. The vibrational energy is given by

$$\varepsilon_n = (n + \tfrac{1}{2})h\nu, \qquad n = 0, 1, 2, \ldots, \quad (10.15)$$

where

$$\nu = \left(\frac{1}{2\pi}\right)\left(\frac{f}{m}\right)^{1/2}, \quad (10.16)$$

for the case of vibration in one dimension.

Similarly we could consider the motion of an electron in the Coulomb potential of an ion core having a charge ze. In this case the potential is

$$U = -\left(\frac{ze^2}{r}\right), \qquad (10.17)$$

and the resulting expression for the energy is

$$\varepsilon = -\left(\frac{mz^2e^4}{8\varepsilon_0^2 h^2 n^2}\right), \qquad n = 1, 2, 3, \ldots. \qquad (10.18)$$

In this equation, e is the electronic charge and ε_0 is the permittivity of a vacuum. The zero of energy is taken as the free electron at infinite distance from the ion core. Thus ε is large and negative for $n = 1$, and it approaches zero for $n \to \infty$. Other appropriate forms of the potential can be used to develop expressions for the energy in other situations.

Note that in all cases of quantum mechanical calculations, the allowed values of the energy will depend on the volume available to the particle; that is, $\varepsilon_i = f(V)$. In general, the smaller the available volume and the lighter the particle, the farther apart the allowed energy levels will be. Note too that, in general, the differences in energy between adjacent vibrational energy levels will be large compared to those associated with rotational motion, which in turn will be large compared to those associated with translational motion.

DEGENERACY AND INTERPARTICLE EFFECTS

In addition to providing us with the means to calculate the energies allowed to particles in any environment, quantum mechanics also introduces two other concepts that we must use in the evaluation of partition functions. The first of these is the concept of the degeneracy of energy levels. The second is the question of the effect that the presence of additional particles has on the energy states available to a given particle.

We have already touched on the question of degeneracy when we talked about the difference between an energy state and an energy level. Quantum mechanical calculations indicate that there are often several allowed quantum states of an individual particle that have the same energy, but differ in some other observable respect—for example, configuration or spin. The importance of degeneracy insofar as evaluating partition functions is concerned is that all of the degenerate states of a given energy level must be counted as separate states when the summing process leading to the partition function is carried out.

The second question, that of interparticle interaction, has two aspects. First, one must consider the effect that the presence of additional particles has on the allowed values of the energy of a given particle. That is, are the solutions to the Schrödinger equation different as a result of the potential field associated with the presence of other particles in the system? The answer to this question, in principle, is yes. The deviation of the pressure in a van der Waals gas from direct proportionality to the number of molecules present is one example of this effect. In practice, however, in many cases it is possible to ignore these interactions, because they introduce changes in the allowed energy values that are small compared to the absolute values of the energies. In other cases, it is possible to consider effects arising from interparticle interactions separately from the energies associated with the various modes of particle motion. In either case, it is often possible to treat the particles that make up the system as though they behaved independently of each other. We shall see shortly that great simplification in the summing process leading to the partition function is possible when this independence can be assumed.

The other aspect of the interparticle interaction problem is concerned with the question of how many particles in a given system can be in a given allowed quantum state at a given time. Quantum mechanics indicates that the allowed number of particles per state depends on the fundamental nature of the particles involved and that the allowed number is either 1 or ∞. All fundamental particles (fundamental from the viewpoint of chemistry or atomic physics) fall into one of two classes:

1. Particles with "antisymmetric" wave functions, which are said to have "half integral spins." These particles follow what is called *Fermi–Dirac statistics* and can have no more than one particle per quantum state. Particles in this class include electrons, protons, neutrons, deuterons, and ^{3}He atoms.
2. Particles with "symmetric" wave functions, which are said to have "integral spins." These particles follow *Bose–Einstein statistics* and can have an infinite number of particles in a given quantum state. Particles in this class include photons, ^{4}He atoms, hydrogen atoms, and hydrogen molecules.

All other atomic and molecular species also fall into one of these two classes. As a practical matter, the effects associated with the distinction between the two types of statistics are observed primarily with very light particles and predominantly, but not always, at low temperatures or high densities. We will see examples of this later. The consequence of the above distinction, insofar as the evaluation of the partition function is concerned, is that this "exclusion principle" must be taken into account when the sum over quantum states is set up. We will find as a practical matter, however, that in many real systems the number of available particle quantum states is very much larger than the number of particles. In this case the probability of finding more than one particle in a given state is very small on a straight statistical basis, and the question of whether to use Fermi–Dirac or Bose–Einstein statistics does not arise. We will see an example of this when we consider the statistical mechanical treatment of the monatomic ideal gas.

INDEPENDENT PARTICLE SYSTEMS

Let us now look in detail at the simplifications in the calculation of the allowed E_i that arise if we can assume that the particles in the system act independently of one another. In developing the equations for the partition functions as sums over available energy states, the only assumption that was made was that we could calculate the appropriate energies, E_i, of the quantum states of the macroscopic system. In general, the specification of a particular E_i would involve specifying the position and the momentum of every particle in the system. In many cases, as mentioned previously, the interactions between particles are sufficiently weak that we can make the simplifying assumption that the energy states available to each particle are independent of the presence of the other particles in the macroscopic system. Moreover, in many cases we can also assume that the various modes of energy storage in a given particle, such as translation, rotation, or vibration, operate independently of one another.

This assumption of independence implies that although the molecules do interact with one another sufficiently strongly to ensure internal equilibrium in the system, the interaction is not strong enough to affect the allowed values of the energies calculated by quantum mechanics. For example, in the case of an ideal gas, the molecules, assumed to be point masses, interact by collision with the walls of the system, but the density is sufficiently low that intermolecular attractive forces are negligible. Similarly, for the case of the internal modes of a given molecule, it can often be assumed that the allowed rotational energy levels are unaffected by the particular vibrational state of the molecule. Such assumptions are always approximations, but in many cases are good enough to permit calculation of useful values of the thermodynamic coordinates of the system.

RELATION OF SYSTEM ENERGY TO PARTICLE ENERGIES

With this concept of independence in mind, let us develop some relations that apply generally to systems of independent particles. First, and most significantly, since we are neglecting interparticle forces, we may state that the energy of a given particle is independent of the presence of other particles, and thus at any instant we have

$$E = \varepsilon_a + \varepsilon_b + \varepsilon_c + \cdots. \tag{10.19}$$

That is, the total energy of the macroscopic system is simply the sum of the energies of the individual particles, and we need not consider interaction energies in order to calculate the energy of the system as a whole.

RELATION OF SYSTEM ENERGY TO PARTICLE ENERGIES 109

Accordingly, we can define, for each particle in the system, a parameter that has the same form as the canonical ensemble partition function for the macroscopic system. That is,

$$q_a = \sum_j e^{-\varepsilon_{a_j}/kT}, \qquad q_b = \sum_j e^{-\varepsilon_{b_j}/kT}, \qquad \text{and so on,} \qquad (10.20)$$

where the summation is over all of the quantum mechanically allowed energy states of the individual particle. Recall that the quantum mechanical calculations made previously showed that, in general, $\varepsilon_a = f(V)$. Since we are considering individual particle energy states, ε_a is independent of N, and the allowed energy states are, in general, independent of T. Thus the ε_a depend only on the type of particle and the system volume. Thus

$$\varepsilon_a = \varepsilon_a(V) \quad \text{and} \quad q = q(V, T). \qquad (10.21)$$

We may now note that taking the product of all the q_a for a given system generates all possible values of E_i—the energy states of the macroscopic system. As an example, let us suppose that we have a system composed of two molecules, a and b, that a has three available energy states a_1, a_2, and a_3, and that b has two available energy states, b_1 and b_2. The product of the q's for this case is

$$q_a \cdot q_b = e^{-\varepsilon_{a_1}/kT} + e^{-\varepsilon_{a_2}/kT} + e^{-\varepsilon_{a_3}/kT} + e^{-\varepsilon_{b_1}/kT} + e^{-\varepsilon_{b_2}/kT}, \qquad (10.22)$$

or

$$q_a \cdot q_b = e^{-(\varepsilon_{a_1}+\varepsilon_{b_1})/kT} + e^{-(\varepsilon_{a_2}+\varepsilon_{b_1})/kT} + e^{-(\varepsilon_{a_3}+\varepsilon_{b_1})/kT}$$
$$+ e^{-(\varepsilon_{a_1}+\varepsilon_{b_2})/kT} + \cdots. \qquad (10.23)$$

Each of the terms in this resulting sum is one of the terms in the partition function for the macroscopic system,

$$Q = \sum_i e^{-E_i/kT}, \qquad (10.24)$$

and the total number of terms in the sum represents all possible values of E_i for the system as defined. Thus

$$Q = q_a \cdot q_b \cdot q_c \cdots. \qquad (10.25)$$

This is a relation of general validity for any system composed of *independent, different, distinguishable* particles.

In spite of the generality of the above expression for Q, it is not a particularly useful expression for treating real systems which are made up of a large number of particles that are either identical or belong to a small number of different species. We must thus modify the above equation to account for the fact that the particles involved are *identical* and may or may not be *distinguishable*.

DISTINGUISHABLE PARTICLE SYSTEMS

Consider first the question of distinguishability. In the present context, this term means that one must be able to identify each particular particle on some physical basis. The classic example of a system of distinguishable particles is a crystalline solid at low enough temperature that diffusion and defect migration cannot take place. In principle, if one had such a crystal, of known geometry, one could count in x atoms from one corner, back y atoms along a face, and down z atoms into the bulk of the crystal and identify a particular atom. If one were to go away, return at some later date, and again count out the same x, y, z coordinates, the atom at this position would be the same identical atom that occupied that site at the previous time. The particles in such a system are thus distinguishable. (It is worth noting that analytical techniques such as scanning tunneling microscopy permit us to carry out this measurement in practice for the atoms at the surface of a material. It is, in fact, possible to monitor the behavior of specific, single atoms over a period of time.)

Alternatively, if one were to consider a system consisting of a large number of particles in a gas phase, there is no way, either in principle or in practice, of keeping track of the position or momentum of a given atom. In this case, one cannot identify a particular particle, go away, return, and have any possibility of identifying that same particle at some later date. The most that one can say in a system of this type is that *some* particle has a given set of position and momentum coordinates. There is no way to tell *which* particle it is that has these coordinates. The particles in this type of system are said to be *indistinguishable*.

With this distinction in mind, let us return to the question of the correct equation for Q in a system of *independent, identical, distinguishable* particles. In this case, since all of the particles are identical, they all have the same available single-particle quantum states. In other words,

$$q_a = q_b = q_c = \cdots = q, \tag{10.26}$$

and, since in a system of N particles there are N terms of type q_a, we have

$$Q = q^N \tag{10.27}$$

for this case of *independent, identical, distinguishable* particles.

INDISTINGUISHABLE PARTICLE SYSTEMS

Now let use consider the case where the particles are not only *independent* and *identical* but *indistinguishable* as well. We can illustrate the problem that arises in this case by returning to the previous example of a system containing two particles, a and b. This time let us assume that the particles are identical and thus have the same available single-particle energy states, say a_1, a_2, a_3 and b_1, b_2, b_3. If we form the product $q_a \cdot q_b$ as before, we have

$$q_a \cdot q_b = e^{-(\varepsilon_{a_1}+\varepsilon_{b_1})/kT} + e^{-(\varepsilon_{a_1}+\varepsilon_{b_2})/kT} + \cdots + e^{-(\varepsilon_{a_2}+\varepsilon_{b_1})/kT} + \cdots. \quad (10.28)$$

If the particles are identical and indistinguishable, then the definition of a state of the system as having particle a in state a_1 and particle b in state b_2 has no meaning. All we can say is that *one* of the particles is in state 1 and *one* of the particles is in state 2. Thus there is no physical distinction between the terms

$$e^{-(\varepsilon_{a_1}+\varepsilon_{b_2})/kT} \quad \text{and} \quad e^{-(\varepsilon_{a_2}+\varepsilon_{b_1})/kT}. \quad (10.29)$$

Since these two cases are not distinguishable on any physical basis, they do not correspond to different states of the macroscopic system. Thus, since the only states that are counted in forming Q as

$$Q = \sum_i e^{-E_i/kT} \quad (10.30)$$

are those that are physically distinguishable from one another, the product

$$q_a \cdot q_b \cdot q_c \cdots \quad (10.31)$$

contains terms that do not belong in Q.

The question that now arises is that of how one can correct the product of the q's to remove the effect of these terms that do not represent different states of the system. In answering this question, it is convenient to separate the terms in the product of q's into terms of two types, namely:

1. Terms in which no two particles are in the same particle quantum state—that is, terms such as

$$e^{-(\varepsilon_{a_1}+\varepsilon_{b_3}+\varepsilon_{c_5}+\varepsilon_{d_2}\cdots)/kT}. \quad (10.32)$$

2. Terms in which at least two particles are in the same particle quantum state—that is,

$$e^{-(\varepsilon_{a_1}+\varepsilon_{b_3}+\varepsilon_{c_3}+\varepsilon_{d_2}\cdots)/kT}. \quad (10.33)$$

112 EVALUATION OF THE ALLOWED ENERGIES

Terms of the first type may be handled straightforwardly. In a system of N particles, each term of type 1 will occur $N!$ times, as $N!$ is the number of ways of arranging the N particles in the N occupied single-particle quantum states. If these were the only type of terms involved, correction of the expression for Q would involve simply division of the expression developed for the case of independent identical distinguishable particles by $N!$.

Terms of type 2 are more difficult to deal with, because they involve the question of the number of particles allowed per quantum state. We touched on this question in the discussion of quantum mechanics earlier in this chapter, when we categorized particles as following Fermi–Dirac statistics, with no more than one particle per quantum state, or Bose–Einstein statistics, with an unlimited number of particles per quantum state. We will not consider situations of these types in detail at this point, simply note that the number of cases in which the question of the number of particles allowed per quantum state has any practical significance are limited, and mostly involve systems of light particles at high densities and at low temperatures. In Part V of this text, we will deal explicitly with two systems in which these considerations *are* important, namely the electron gas and the photon gas, or blackbody radiation.

Fortunately, there is another way around the problem of the terms with more than one particle per quantum state that takes care of most situations of practical interest. This is simply that, in very many systems made up of indistinguishable particles, the number of available single-particle quantum states is *very large* compared to the number of particles in the system. We will prove, for example, that in the case of an ideal gas the number of translational energy state having energies on the order of kT or less, for ordinary temperatures, is extremely large compared to the number of molecules in a mole of gas. The practical consequence of this is that the *probability* of finding two particles in the same state is extremely small, and thus such terms make negligible contribution to the product of the q's and may be safely ignored. Systems in which this behavior is observed are said to follow *Boltzmann statistics*. For such systems the only correction required is division by $N!$. Thus, in this case,

$$Q = \frac{q^N}{N!} \qquad (10.34)$$

for a system of *independent, identical, indistinguishable* particles following Boltzmann statistics.

INDEPENDENCE OF MODES OF ENERGY STORAGE

In many systems, we can make still one more approximation that will simplify the process of evaluating the partition function—namely, that not only are the energies of the individual particles that make up the system independent of one another, but the various modes of energy storage available to a given particle are essentially independent of one another. For example, a diatomic molecule may possess translational kinetic energy, energy due to rotation of the molecule about its center of mass, and energy due to vibrational motion of one atom relative to the other. To a first approximation, we can separate these contributions and write that

$$\varepsilon_a = \varepsilon_{a_{\text{trans}}} + \varepsilon_{a_{\text{rot}}} + \varepsilon_{a_{\text{vib}}} + \cdots. \quad (10.35)$$

If we define terms such as

$$q_{\text{trans}} = \sum_n e^{-\varepsilon_n/kT}, \quad (10.36)$$

where the sum is taken over all of the allowed translational energy states, and realize that

$$q_a = \sum_i e^{-\varepsilon_{a\text{trans}} + \varepsilon_{a\text{rot}} + \varepsilon_{a\text{vib}} + \cdots /kT}, \quad (10.37)$$

we may write

$$q_a = q_{\text{trans}} \cdot q_{\text{rot}} \cdot q_{\text{vib}} \cdots. \quad (10.38)$$

This expression may be substituted into the expression for Q for the case of independent, identical, indistinguishable particles to yield

$$Q = \left(\frac{1}{N!}\right)(q_{\text{trans}} \cdot q_{\text{rot}} \cdot q_{\text{vib}} \cdots)^N \quad (10.39)$$

as the final expression for Q for a gas of diatomic molecules.

BIBLIOGRAPHY

T. L. Hill, *Introduction to Statistical Thermodynamics*, Addison-Wesley, 1960, Chapter 3.

C. C. Kittel and H. Kroemer, *Thermal Physics*, 2nd ed., W. H. Freeman & Co., San Francisco, 1980, Chapters 1 and 6.

J. H. Knox, *Molecular Thermodynamics*, John Wiley & Sons, New York, 1978, Chapters 2 and 5.

E. L. Knuth, *Introduction to Statistical Thermodynamics*, McGraw-Hill, New York, 1966, Chapters 4, 6, and 7.

W. G. V. Rosser, *An Introduction to Statistical Physics*, Ellis Horwood, Chichester, UK, 1982, Chapter 2.

PROBLEMS

10.1 What is the wavelength of an electron whose kinetic energy is 100 electron volts? Does this result have any relevance to the possible resolving power of an electron microscope? Would the resolving power, in theory, be improved by using protons instead of electrons?

10.2 Consider an $O^{16}O^{16}$ oxygen molecule having a mass of exactly 32 daltons. Imagine it to be contained in a 1-cm cube at 300 K. What is its average translational kinetic energy, as given by $\varepsilon = \frac{3}{2}kT$? Assuming that the components of its momentum in all three directions in space are equal, what must p_x, p_y, and p_z be in order that the molecule shall possess the average translational energy? What must the translational quantum numbers be to give this energy?

10.3 For a gas of independent molecules,

$$Q = \frac{[q(V, T)]^N}{N!}.$$

From the canonical equations

$$\mu = \frac{G}{N} = -kT\left(\frac{\partial \ln Q}{\partial N}\right)_{V,T} \quad \text{and} \quad P = kT\left(\frac{\partial \ln Q}{\partial V}\right)_{N,T}$$

prove that $q = Vf(T)$.

10.4 Extend the derivation of the equation

$$Q = \frac{q^N}{N!}$$

for a one-component system of independent, indistinguishable, identical molecules to the case of a two-component system, for which

$$Q = \frac{q_1^{N_1} \cdot q_2^{N_2}}{(N_1)!(N_2)!}.$$

10.5 Consider a system made up of four identical particles in a container of volume V. Assume that each particle has available to it two energy states, ε_1 and ε_2, with $\varepsilon_1 < \varepsilon_2$. Assume that the particles are bosons (i.e., there is no limit to the number of particles in the system in the same quantum state).

(a) Write an expression for q for a single particle.
(b) Write an expression for Q for the system, assuming the particles to be distinguishable.

(c) Write an expression for C_V for the system, and show that $C_V \to 0$ as $T \to 0$, $C_V \to 0$ as $T \to 0$, and C_V has a maximum at some finite temperature.

10.6 Consider a system consisting of N distinguishable particles, each of which has two available energy states, $\varepsilon_1 = 1k$, $\varepsilon_2 = 3k$. Calculate, using statistical thermodynamics, the energy, entropy, and heat capacity at constant volume for this system at $T = 2/k$ K. Assume that the particles are bosons, with an infinite number of particles allowed per energy state.

10.7 Consider a system consisting of a gas of independent, identical, indistinguishable particles. The only limitation on the number of particles in the system is that the particles are fermions, and the system can thus have no more than one particle per state. The energy states available to the particles are $\varepsilon_1 = 1$, $\varepsilon_2 = 1$, and $\varepsilon_3 = 3$.

(a) What kind of ensemble would be appropriate to describe this system?

(b) List the available system microstates, in terms of the values of N and $E_i(N)$ for each state.

(c) If $T = 1/k$, $\mu = 2$, and $V = 1$ cm^3, what is the pressure in the system?

PART III

SINGLE-COMPONENT SYSTEMS

CHAPTER 11

CLASSICAL THERMODYNAMICS OF ONE-COMPONENT SYSTEMS

We are now in a position to develop thermodynamic expressions to describe the behavior of real systems. The order of presentation in this description will be first to develop the appropriate relations for each class of systems on a classical basis, and then to use statistical mechanics to evaluate the terms that appear in the relations derived classically. We will analyze the behavior of systems primarily in terms of the Gibbs free energy, G, because in most cases of practical interest in materials science the independent variables will be pressure and temperature. A minimum in the Gibbs free energy, as you will recall from the treatment in Chapter 3, is the criterion for equilibrium in a system of constant pressure and temperature.

We will begin with the consideration of one-component systems, because these are the simplest. We will then go on an consider systems of more than one component, as well as systems in which chemical reactions can take place. Finally, we may consider the complications that arise when such things as interfaces and defects contribute significantly to the overall properties and behavior of the system.

FREE ENERGY SURFACES

Let us begin by considering a single phase in internal equilibrium in a one-component system. For such a phase, we can write an expression for the

Gibbs free energy at any value of P and T as

$$G^\alpha(P, T) = G_0^\alpha(P_0, T_0) + \int_{P_0}^{P} \left(\frac{\partial G^\alpha}{\partial P}\right)_T dP + \int_{T_0}^{T} \left(\frac{\partial G^\alpha}{\partial T}\right)_P dT, \quad (11.1)$$

in which $G_0^\alpha(P_0, T_0)$ is an arbitrarily chosen standard-state Gibbs free energy. We know from the treatment of classical thermodynamics in Chapter 4 that

$$\left(\frac{\partial G}{\partial T}\right)_P = -S, \quad \left(\frac{\partial^2 G}{\partial T^2}\right)_P = -\left(\frac{\partial S}{\partial T}\right)_P = -\frac{C_p}{T} \quad (11.2)$$

and

$$\left(\frac{\partial G}{\partial P}\right)_T = V, \quad \left(\frac{\partial^2 G}{\partial P^2}\right) = \left(\frac{\partial V}{\partial P}\right)_T = -\kappa V. \quad (11.3)$$

Thus we see that, in principle, once a value for $G_0^\alpha(P_0, T_0)$ is chosen, we may calculate the value of G^α at any other value of P and T from a knowledge of parameters such as S, V, κ, and C_P. These values may in turn be obtained classically from experimental measurements in the system of interest, or, statistically, by first evaluating the partition function for the system on the basis of a model on the molecular scale and then using the previously developed relations between the thermodynamic state functions and the partition function.

In either case, the results of the calculation can be represented graphically in a space whose axes are G, P, and T by a surface—a so-called free energy surface—as shown schematically in Figure 11.1.

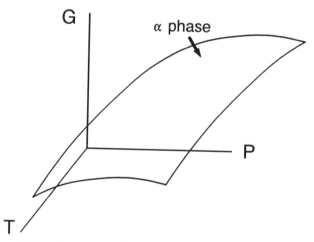

Figure 11.1 A Gibbs free energy $(G-P-T)$ surface for a single phase of a one-component system in internal equilibrium.

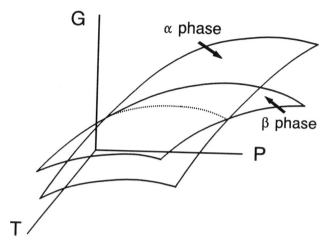

Figure 11.2 Gibbs free energy (G–P–T) surfaces for two phases in a one-component system. The two phases are in mutual equilibrium along the dotted line where the two surfaces intersect.

Similarly one can write an expression for the Gibbs free energy as a function of pressure and temperature for any other phase in this same system, referred to as the β phase:

$$G^\beta = G_0^\alpha(P_0, T_0) + \Delta G_0^{\alpha\beta}(P_0, T_0) + \int_{P_0}^{P} \left(\frac{\partial G^\beta}{\partial P}\right)_T dP + \int_{T_0}^{T} \left(\frac{\partial G^\beta}{\partial T}\right)_P dT, \quad (11.4)$$

where $\Delta G_0^{\alpha\beta}$ is the difference between G^α and G^β at T_0, P_0 and is given by

$$\Delta G_0^{\alpha\beta}(P_0, T_0) = \Delta G^{\alpha\beta}(P_0, T_{eq}) - \int_{T_{eq}}^{T_0} \Delta S^{\alpha\beta}(P_0, T) \, dT, \quad (11.5)$$

or

$$\Delta G_0^{\alpha\beta}(P_0, T_0) = -\int_{T_{eq}}^{T_0} \Delta S^{\alpha\beta}(P_0, T) \, dT, \quad (11.6)$$

where T_{eq} is the temperature at which the α and β phases are in equilibrium when the pressure is P_0. The term $\Delta G^{\alpha\beta}(P_0, T_{eq})$ is equal to zero, as the criterion for equilibrium is that $G^\alpha = G^\beta$ at T_{eq}. Since the expression for G^β is referenced to the expression for G^α, a free energy surface for the β phase can also be drawn in the same G, P, T space used before, as shown in Figure 11.2. Similar expressions for G, and similar free energy surfaces, could be drawn for all other conceivable phases of the system in question.

TEMPERATURE DEPENDENCE OF THE THERMODYNAMIC FUNCTIONS

Let us now spend some time considering the general shape of the free energy surfaces defined above. Because of the complexity of these surfaces and also the difficulty of representing a three-dimensional surface in two dimensions, we will consider the behavior of G as a function of P and T separately. That is, we will look at a cut through the G–P–T surfaces at a constant P or constant T. In such a case we will be looking at a two-dimensional plot which will have on it lines representing the intersections of the free energy surfaces with planes of constant P or T.

Let us look first at a plot of G versus T on a plane of constant P. Such a plot is shown in Figure 11.3, where we have assumed that the only stable phases present in the system studied are a solid, a liquid, and a vapor. The general shape of the curves on this plot can be obtained from consideration of the partial derivatives of G with respect to T. Since

$$\left(\frac{\partial G}{\partial T}\right)_P = -S$$

and we know that S is always positive, the slopes of the curves will be negative. Moreover, it is generally observed for most systems that

$$S_s < S_l \ll S_v, \tag{11.7}$$

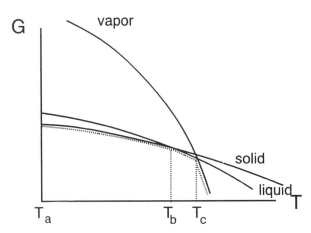

Figure 11.3 A constant pressure cut through a G–P–T plot for a typical one-component system. The phase present at equilibrium at any temperature for this pressure is shown by the dotted line.

so that the slopes of the curves will also be in this same order. Moreover,

$$\left(\frac{\partial^2 G}{\partial T^2}\right)_P = -\frac{C_p}{T},$$

and again, since C_P and T are both always positive, the curvature of the curves shown will be negative. Finally, since

$$G = H - TS,$$

$G \to H$ as $T \to 0$. Since one observes in general that

$$H_s < H_l \ll H_v, \tag{11.8}$$

the intersection of the curves shown with the $T = 0$ axis is as drawn on the diagram.

Note that since the criterion for equilibrium in the system is that G be a minimum, the stable regions for each phase at any P and T are given by the lower envelope of the curves drawn. That is, from T_a to T_b the solid phase is stable, at T_b solid and liquid coexist, from T_b to T_c the liquid phase is stable, at T_c liquid and vapor coexist, and at $T > T_c$ the vapor phase is stable. Keep in mind that this diagram was drawn for some particular value of P. At other values of P the relative positions of the curves will differ. At some values of P, one or more of the phases may have *no* range of stability.

We may also work from this curve to look at the variation of S with temperature—that is, the slope of the G versus T curve. We will look at this only for the stable portions of the G versus T curves; that is, we will plot S versus T as we traverse the lower envelope of the G versus T curves. The plot so obtained will look generally like the one shown in Figure 11.4.

The value of S starts at some low value, S_0, at $T = 0$ K. Through the range of stability of the solid phase, S increases, with the slope of the curve being C_P/T for the solid phase. At the equilibrium temperature between solid and liquid, where there is a discontinuity in the slope of the stable G versus T curve, there is a finite discontinuity in S, namely ΔS_f, the entropy of fusion. Above this temperature, S again rises with T, with a slope given by C_P/T for the liquid phase. At the equilibrium temperature between liquid and vapor, there is again a discontinuity in S. In this case, it is ΔS_v, the entropy of vaporization.

Again note that this curve is drawn for some particular value of P. The major change that one would see for curves representing other pressures is that as P is increased, the magnitude of ΔS_v would decrease, and eventually go to zero at the critical pressure of the system, as shown by the dotted curves on Figure 11.4.

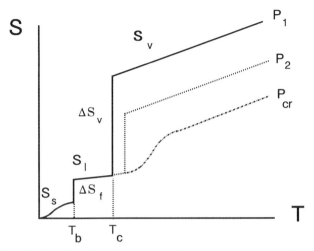

Figure 11.4 A plot showing the entropy, S, versus temperature behavior for a typical one-component system. The discontinuity in S between liquid and vapor phases disappears at P_{cr}.

As a final point in our consideration of the relationship between G and T, we may plot G/T versus $1/T$. Such a plot is shown in Figure 11.5. In this case, we can deduce the general form of the relationships involved using

$$\left[\frac{\partial(G/T)}{\partial(1/T)}\right] = H, \qquad \left(\frac{\partial H}{\partial T}\right) = C_P. \tag{11.9}$$

Since both H and C_P are positive, we expect in this case that the slopes of the curves will be positive and the curvatures will be negative. Also, since

$$H_s < H_l \ll H_v,$$

we see this reflected in the magnitudes of the slopes and the values of G/T at low temperatures (large values of $1/T$) for the various phases. Again as in the plot of G versus T, the lower envelope of the curves defines the stable phases at various values of $1/T$.

In this case we can also plot the slopes of the stable parts of the curve against temperature, as shown in Figure 11.6. Here H starts at a low value, H_0, for the solid phase, and it increases with temperature with a slope equal to C_P for the solid. At the equilibrium temperature between solid and liquid, there is a discontinuous change in H representing the enthalpy of fusion, ΔH_f. Above this temperature the value of H again rises with temperature, with a slope of C_P for the liquid phase. Again at the equilibrium temperature between liquid

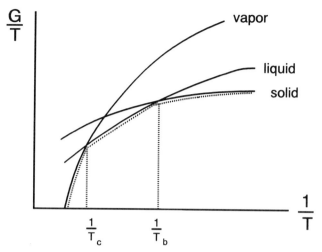

Figure 11.5 A plot showing the behavior of G/T versus $1/T$, at constant pressure for a typical one-component system. The slope of the curves at any point is the enthalpy, H, of that phase. The dotted line indicates the stable phase at any value of $1/T$.

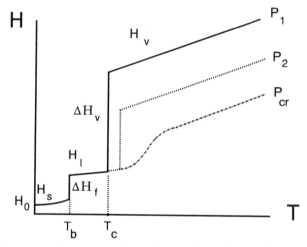

Figure 11.6 A plot showing the behavior of the enthalpy, H, versus temperature at constant pressure for a typical one-component system. The discontinuity in H between liquid and vapor disappears at P_{cr}.

and vapor there is a discontinuity in H, in this case ΔH_v, the enthalpy of vaporization. As was the case for ΔS_v in the graph of S versus T, the value of ΔH_v decreases with increasing pressure, and the finite discontinuity disappears at the critical pressure. This behavior is shown by the dotted curves in Figure 11.6.

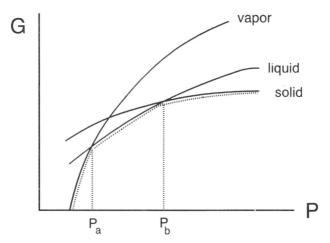

Figure 11.7 A constant temperature cut through the $G-P-T$ plot for a typical one-component system. The phases present at equilibrium at any pressure for this temperature are shown by the dotted line.

PRESSURE DEPENDENCE OF THE THERMODYNAMIC FUNCTIONS

We may also investigate the dependence of G on P at constant temperature by looking at a cut through the $G-P-T$ space at constant T. Such a plot is shown schematically in Figure 11.7. In this case we know that

$$\left(\frac{\partial G}{\partial P}\right)_T = V, \qquad \left(\frac{\partial^2 G}{\partial P^2}\right)_T = -\kappa V.$$

Since V and κ are both always positive, the slopes of the curves in this case will be positive and the curvatures will be negative. In most systems we have

$$\mathbf{V}_s < \mathbf{V}_l \ll \mathbf{V}_v, \tag{11.10}$$

where $\mathbf{V}_i$ is the volume per mole of each phase. Thus the slopes of the lines representing the three phases will differ correspondingly. Again the lower envelope of the curves defines the regions of stability of the various phases, with vapor being stable below P_a, vapor and liquid coexisting at P_a, liquid being stable between P_a and P_b, liquid and solid coexisting at P_b, and solid being stable above P_b. Again, this represents the situation at some particular value of T, and the relative positions of the lines would change with changing T.

We may also look at the variation of V with P by plotting the slopes of the portions of the G versus P curves representing the regions of stability of the

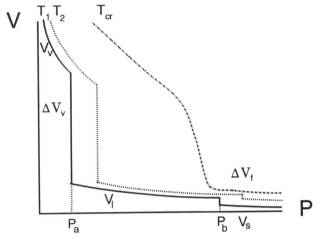

Figure 11.8 A plot showing the variation of volume, V, with pressure in a typical one-component system. The discontinuity in V between liquid and vapor disappears at T_{cr}.

various phases. Such a plot is shown in Figure 11.8. At low pressures the volume per mole is high, and it drops rapidly with increasing pressure due to the high compressibility of the vapor phase. At the vapor–liquid equilibrium pressure, P_a, there is a discontinuous change in V, which is ΔV_v, the volume change on vaporization. As the pressure is further increased, V decreases further, but at a much slower rate, due to the much smaller compressibility of the liquid. At the liquid–solid equilibrium pressure, P_b, there is another discontinuous change in V, in this case ΔV_f, the volume change on fusion. As the pressure is further increased, there is a further decrease in V associated with the small but finite compressibility of the solid. At higher temperatures the situation would be similar, but the curves would be displaced to higher values of pressure, and the discontinuity in volume at the vapor–liquid equilibrium temperature would decrease, and finally go to zero at the critical temperature, as shown by the dotted curves in Figure 11.8.

THE ONE-COMPONENT PHASE DIAGRAM

As a final way of looking at the implications of the G–P–T surfaces, we may consider the plot that results when we project the intersections of the surfaces representing the various phases onto the P–T plane. Note that this is a different process than we have been considering heretofore. This is *not* a cut at constant G, but a projection of the lines of intersection onto the P–T plane, as

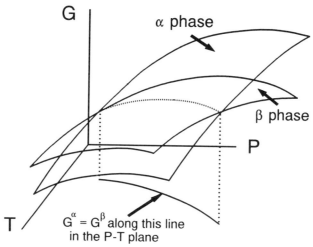

Figure 11.9 Gibbs free energy (G–P–T) surfaces for two phases in a one-component system, showing the line of intersection of the two surfaces projected onto the P–T plane.

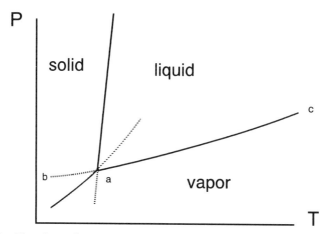

Figure 11.10 The phase diagram for a typical one-component system. Two phases are in equilibrium along each of the curves shown (coexistence curves). Point c represents the critical point, where the distinction between liquid and vapor disappears. Point a is the triple point, at which all three phases are in mutual equilibrium. The dotted line segments, such as ab, represent metastable equilibrium between two phases.

shown schematically in Figure 11.9. The resulting plot, which is the one-component equilibrium phase diagram, is shown in Figure 11.10. Areas on this figure represent conditions of P and T in which a single phase is stable. Lines represent values of P and T at which two free energy surfaces intersect. Along this line both phases have the same G and, consequently, are in equilibrium.

At the point of intersection of the three lines, all three phases coexist at equilibrium. The dotted portions of the lines represent metastable equilibrium points. For example, vapor and liquid would be in equilibrium with each other, but have a higher Gibbs free energy than the solid phase along the line segment *ab*. Note that there can be no point in a one-component system where the free energy surfaces of four phases coincide. This is a consequence of the Gibbs phase rule, which we derived in Chapter 5 as

$$DOF = C + 2 - P.$$

Since the number of DOF can never be negative, the maximum number of phases that can coexist in a one-component system is three.

The only other point of importance on the one-component phase diagram is point c, at the termination of the liquid–vapor coexistence line. This represents the critical point, which is the point at which the parameters such as ΔS_v, ΔH_v, and ΔV_v go to zero, as shown in the previous figures, and the distinction between the vapor and liquid phases disappears.

MOLAR PROPERTIES

Let us consider in more detail the relationships between P and T along the coexistence lines. Before we do this, however, let us introduce an additional concept, namely, the concept of molar properties. We shall see that this concept is useful, in that it allows us to treat the extensive thermodynamic properties as though they were intensive properties, and allows us to express the properties of any system on a *per mole* basis. We begin by defining the mole fraction, X_i, of a given component i as

$$X_i = \frac{n_i}{\sum_i n_i} \qquad (11.11)$$

in which n_i is the number of moles of component i in the system and $\sum_i n_i$ is the total number of moles of *all* components in the system. We can similarly define the other extensive thermodynamic state functions on a molar basis. For example, we may define the molar entropy as

$$\mathbf{S} = \frac{S}{\sum_i n_i}, \qquad (11.12)$$

and similarly for the other state functions such as V, H, G, F, and E. (Note that we will use bold symbols for molar properties throughout this text.)

THE CLAPEYRON EQUATION

With this concept in mind, let us now consider the changes that take place when we carry out an infinitesimal process about any value of P and T which lies on one of the coexistence lines on the one-component phase diagram. Since we are starting at a point where two phases are in equilibrium, we may write one Gibbs–Duhem equation for the changes taking place in each of the two phases during the infinitesimal process:

$$\sum_i X_i^\alpha N_{av} d\mu_i^\alpha + S^\alpha dT - V^\alpha dP = 0, \tag{11.13}$$

$$\sum_i X_i^\beta N_{av} d\mu_i^\beta + S^\beta dT - V^\beta dP = 0, \tag{11.14}$$

where we have written the Gibbs–Duhem equations in terms of the molar properties defined above. If we now consider the form of these equations for the special case where the infinitesimal process is carried out *along* the coexistence line, we may write that

$$d\mu_i^\alpha = d\mu_i^\beta, \tag{11.15}$$

as the two phases are in equilibrium throughout the process. Moreover, since we are dealing with a one-component system, we have

$$X_i^\alpha = X_i^\beta = 1. \tag{11.16}$$

Thus

$$S^\alpha dT - V^\alpha dP = S^\beta dT - V^\beta dP, \tag{11.17}$$

or

$$\frac{dP}{dT} = \frac{\Delta S}{\Delta V}, \tag{11.18}$$

where

$$\Delta S = S^\alpha - S^\beta, \qquad \Delta V = V^\alpha - V^\beta.$$

Moreover, since we know that

$$\Delta G = \Delta H - T\Delta S = 0 \tag{11.19}$$

for this process taking place at equilibrium about a given P and T, we may substitute that

$$\Delta S = \frac{\Delta H}{T} \tag{11.20}$$

to yield

$$\frac{dP}{dT} = \frac{\Delta H}{T \Delta V}. \tag{11.21}$$

Both equations (11.18) and (11.21) are known as the *Clapeyron equation*, giving the slope of the coexistence curve between two phases. Integration of this expression over appropriate limits gives the equation of the coexistence line.

Let us look at application of the Clayeyron equation to two typical cases. Consider first the coexistence line between two condensed phases, say solid and liquid. In this case the most convenient form of the equation to use is

$$\frac{dP}{dT} = \frac{\Delta S_f}{\Delta V_f}.$$

Values of ΔS_f and ΔV_f for a number of materials are given in Appendix C. It can be seen from this list that, for many materials, ΔS_f is on the order of 10 J/mol-K, and the change in molar volume on melting is typically less than 1 cm^3/mol. Putting these numbers into the above equation and using the appropriate conversion factor yields

$$\frac{dP}{dT} \approx 100 \text{ atm/K}. \tag{11.22}$$

That is, the slope of the coexistence line between two condensed phases is very steep. Note that if ΔV_f is positive (that is, if the substance expands on melting), dP/dT will be positive. This behavior is typical of systems that have close-packed structures in the solid state. If ΔV_f is negative, as it is for substances that contract on melting, dP/dT will be negative. This behavior is typical of substances (such as water or bismuth) that have non-close-packed structures in the solid state.

Alternatively, if we look at the coexistence line between condensed and vapor phases, the more convenient form of the Clapeyron equation is

$$\frac{dP}{dT} = \frac{\Delta H_v}{T \Delta V_v}. \tag{11.23}$$

In this case

$$\Delta V = V_v - V_c \approx V_v \approx \frac{RT}{P}, \qquad (11.24)$$

if we assume the vapor phase to be an ideal gas. This leads to

$$\frac{dP}{dT} = \frac{\Delta H_v P}{RT^2}, \qquad (11.25)$$

or

$$\frac{d(\ln P)}{d(1/T)} = -\frac{\Delta H_v}{R}. \qquad (11.26)$$

This last equation is the familiar form of the relation used to describe the temperature variation of the equilibrium vapor pressure over a condensed phase. A typical plot of ln P versus $1/T$ is shown in Figure 11.11, and empirical vapor pressure versus temperature relations are summarized for a number of substances in Appendix D. Note that if ΔH_v is independent of temperature, the curve shown in Figure 11.11 will be a straight line, otherwise it will exhibit some curvature, as is shown in the figure.

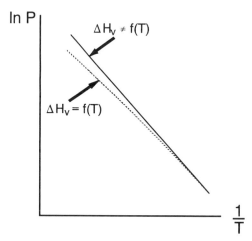

Figure 11.11 The equilibrium vapor pressure curve for a one-component system. The solid curve represents a system for which the enthalpy of vaporization, ΔH_v, is independent of temperature. The dotted curve allows for a reduction in ΔH_v with increasing temperature, the commonly observed case.

EVALUATION OF THE STATE FUNCTIONS

We have thus shown that we can describe the phase relations in a one-component system in terms of the thermodynamic state functions for the various phases as a function of T and P. What is needed now is a means of evaluating these state functions in a particular system—that is, to determine H, S, G, and the equation of state for each phase. One can approach this problem in either of two ways. On the one hand, the problem can be treated using classical thermodynamics, using the equations that we have developed to describe the state functions in terms of observables, and experimental measurements of such properties as α, κ, C_P, and C_V. Alternatively, we can make a statistical mechanical calculation to evaluate a partition function such as Q or Ξ on the basis of a model developed to describe the system of interest, and then use the relations developed for the state functions in terms of the partition function to evaluate the desired thermodynamic functions.

In the chapters that follow, we will develop the latter approach for a number of models of gaseous, liquid, and solid phases of pure materials.

BIBLIOGRAPHY

R. T. DeHoff, *Thermodynamics in Materials Science*, McGraw-Hill, New York, 1993, Chapter 7.

D. R. Gaskill, *Introduction to Metallurgical Thermodynamics*, McGraw-Hill, New York, 1973, Chapter 7.

P. Gordon, *Principles of Phase Diagrams in Materials Systems*, McGraw-Hill, New York, 1968, Chapter 3.

D. V. Ragone, *Thermodynamics of Materials*, Volume 1, John Wiley & Sons, New York, 1995, Chapter 8.

PROBLEMS

11.1 Consider a vacuum furnace containing a crucible of magnesium. What is the value of the best vacuum attainable at 1300 K in the system? You may assume that the vaporization rate is fast enough to keep up with the reduction in pressure by the vacuum pumps.

11.2 Calculate the heat of sublimation of BeO(s) at 300 K and at 1000 K from vapor pressure data. Estimate the value at 0 K.

11.3 At the melting point, would ΔH_s become discontinuous? What is the relation between ΔH_v and ΔH_s of liquid and solid, respectively.

11.4 Calculate the vapor pressure at 300 K of an element for which the melting point is 1000 K, the normal boiling point is 2500 K, the heat of vaporization is 240,000 J/mol, and the heat of fusion is 12,000 J/mol.

11.5 Determine the melting point at 1000 MPa of a substance that has the following properties at 0.1 MPa:

$T_M = 1000\,\text{K},$ $\qquad \Delta H_f = 12{,}000\,\text{J/mol},$

$V_s = 10.5\,\text{cm}^3/\text{mol},$ $\qquad V_l = 11.0\,\text{cm}^3/\text{mol},$

$\kappa_s = 3 \times 10^{-6}\,\text{MPa}^{-1},$ $\qquad \kappa_l = 4 \times 10^{-6}\,\text{MPa}^{-1}.$

11.6 Prove that the slope of the sublimation curve is steeper than the slope fo the vaporization curve at the triple point of a one-component system.

CHAPTER 12

THE MONATOMIC IDEAL GAS

The first system that we will treat using statistical mechanics is the ideal monatomic gas. The statistical thermodynamic definition of an ideal gas is one in which the particles behave as noninteracting point masses.

THE MODEL

The model in this case is one in which it is assumed that the atoms in the gas do not interact energetically with one another, and that these atoms do not have a finite volume. Equilibrium in the gas is maintained by collisions with the walls of the container confining the gas. In statistical mechanical terms, this means that the particles can be treated as *independent*. Since for a pure material all of the particles are some one kind of atom, they are also *identical*, and since we are talking about a gas they are *indistinguishable*. The appropriate expression for the partition function will thus be

$$Q = \frac{q^N}{N!}, \qquad (12.1)$$

provided we can assure ourselves that the conditions required for Boltzmann statistics are met. That is, the number of quantum states available to each particle must be large compared to the number of particles. This condition must be satisfied in order for $1/N!$ to be the appropriate factor to correct for overcounting of states.

NUMBER OF AVAILABLE STATES

The only contributions to the thermodynamic functions of a monatomic ideal gas will arise from the translation of the atoms within the available volume, V, and the excitation of electronic excited states of the atom. In order to justify the use of the above expression for Q, it will suffice to show that the number of states available for *any one* of the modes is large compared to the number of particles. We will show that the number of available translational energy states is sufficiently large to meet this requirement.

The allowed translational energy states of an atom moving in a box of volume $V = L^3$ are given, as we recall from the brief treatment of quantum mechanics presented in Chapter 10, by

$$\varepsilon_{l,m,n} = \left(\frac{h^2}{8mL^2}\right)(l^2 + m^2 + n^2), \qquad l, m, n = 1, 2, 3, \ldots \qquad (12.2)$$

Each atom will possess three degrees of translational freedom, represented by the three independent quantum numbers l, m, and n. There is no inherent degeneracy in these states as long as l, m, and n are considered to be separately identifiable.

In order to be counted as available in determining the relation between the number of molecules and the number of available states, an energy state must have a significant probability of occupation. In other words, $\exp(-\varepsilon/kT)$ for this state must have an appreciable value, or alternatively ε must not be much greater than kT. Consequently if we can show that the number of states with $\varepsilon \leqslant kT$ is large compared to the number of atoms, the required criterion for the use of Boltzmann statistics will have been met.

In order to make the required calculation, we must have some way of arranging the allowed states according to their energy. We can do this using a three-dimensional quantum number space. Each allowed l–m–n combination will be represented by a point in this quantum number space. Such a space is shown in Figure 12.1. Note that points will appear only in the octant of this space for which l, m, and n are all positive, because these are the only quantum mechanically allowed values of l, m and n. From the equation for $\varepsilon_{l,m,n}$ it can be seen that states of low energy lie close to the origin in this space, with the energy increasing radially outward from the origin. Keep in mind, too, that a volume in this space is not a physical volume (e.g., cm^3) but is, instead, a number of points representing different l–m–n combinations.

The l, m, and n points corresponding to states having an energy less than some specified value, ε, will lie in the positive octant of a sphere in this quantum number space having a radius defined, from the geometry of the system, by

$$R^2 = (l^2 + m^2 + n^2). \qquad (12.3)$$

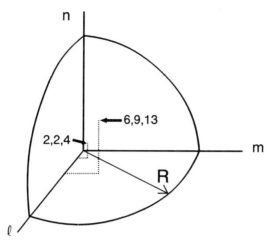

Figure 12.1 A three-dimensional quantum number space, constructed to represent the allowed translational energy states of an atom in a monatomic ideal gas. Each allowed state is represented by a point in this space. The radius, R, defines a volume in this space enclosing the points representing states having an energy less than or equal to the value given in equation (12.7).

Since we know that

$$\varepsilon = \left(\frac{h^2}{8mL^2}\right)(l^2 + m^2 + n^2) \tag{12.4}$$

we may write that

$$(l^2 + m^2 + n^2) = \frac{8mL^2\varepsilon}{h^2} = \frac{8mV^{2/3}\varepsilon}{h^2}. \tag{12.5}$$

Thus the desired sphere radius is

$$R = \left(\frac{8mV^{2/3}\varepsilon}{h^2}\right)^{1/2}, \tag{12.6}$$

and the appropriate volume, which we will call $G(\varepsilon)$ and which enumerates the allowed translational energy states having energy less than or equal to ε, is

$$G(\varepsilon) = \left(\frac{4\pi R^3}{3}\right)\left(\frac{1}{8}\right), \tag{12.7}$$

or

$$G(\varepsilon) = \frac{\pi}{6}\left(\frac{8m\varepsilon}{h^2}\right)^{3/2} V. \qquad (12.8)$$

We are interested in determining the value of $G(\varepsilon)$ for $\varepsilon = kT$. This is

$$G(kT) = \frac{\pi}{6}\left(\frac{8mkT}{h^2}\right)^{3/2} V. \qquad (12.9)$$

For typical gaseous systems, V will be on the order of 1000 cm³, T will be on the order of 300 K, and m will be on the order of 10^{-22} gram. Substituting these values into the expression for $G(kT)$ yields

$$G(kT) \approx 10^{33}. \qquad (12.10)$$

At the volume and temperature given above, the system would contain on the order of 10^{22} particles at atmospheric pressure. Thus the ratio of available states to particles in the system is on the order of 10^{11}, which is certainly large compared to unity. We are thus justified in using the expression

$$Q = \frac{q^N}{N!} \qquad (12.11)$$

to calculate the properties of the monatomic ideal gas.

Before we leave the calculation of the number of available states, note in passing the dependence of $G(kT)$ on the physical parameters of the system. $G(kT)$ would be relatively small at high densities (small values of V/N), at low temperatures, and for small masses. It is in systems like this—for example, helium gas at high pressure and low temperature—that the approximation that $G(kT)$ is large compared to N breaks down, and specific account must be taken of whether the particles involved follow Bose–Einstein or Fermi–Dirac statistics.

EVALUATION OF q

We may now proceed with the evaluation of q. The only modes of energy storage available to the particles in a monatomic ideal gas are those associated with translation of the particles in space and electronic excitation of the atom. (We will not consider the possibility of excitation of the atomic nucleus, because the energies involved in this process are so large as to make the probability of nuclear excitation vanishingly small in any system that we will consider.) We may consider that the translational and electronic excitation modes are independent of one another, and thus write

$$q = q_t \cdot q_e, \tag{12.12}$$

where the subscripts t and e stand for the translational and electronic contributions to q, respectively. We will proceed first to evaluate q_t, because, as we shall see, the major aspects of the behavior of the monatomic ideal gas can be deduced from a knowledge of this parameter alone. Evaluation of q_t means, in principle, evaluation of the sum

$$q_t = \sum_{l,m,n=1}^{\infty} e^{\varepsilon_{l,m,n}/kT}. \tag{12.13}$$

We know from our previous derivation of $G(\varepsilon)$ that a very large number of terms contribute to this sum. Thus it would be very convenient if we could rewrite the sum in terms of an integral and carry out the required integration. We will be justified in doing this if the energy increment between adjacent translational energy states, $\Delta\varepsilon$, is small compared to kT, or if $\Delta\varepsilon/kT \ll 1$. From the expression for $\varepsilon_{l,m,n}$, we see that the smallest possible change in ε is represented by a change in the sum $(l^2 + m^2 + n^2)$ of one unit. Thus

$$\Delta\varepsilon = \frac{h^2}{8mV^{2/3}}, \tag{12.14}$$

or

$$\frac{\Delta\varepsilon}{kT} = \frac{h^2}{8mkTV^{2/3}}. \tag{12.15}$$

Again using $V = 1000\,\text{cm}^3$, $T = 300\,\text{K}$, and $m = 10^{-22}$ gram, we have

$$\frac{\Delta\varepsilon}{kT} \approx 10^{-21}, \tag{12.16}$$

which is obviously very small compared to unity. We may thus set up the sum for q_t as an integral. This means essentially making the substitution

$$q_t = \sum_{l,m,n=1}^{\infty} e^{-\varepsilon_{l,m,n}/kT} = \int_0^{\infty} g(\varepsilon) e^{-\varepsilon/kT}\, d\varepsilon, \tag{12.17}$$

where $g(\varepsilon)$ is the number of states in the energy increment $d\varepsilon$ at any particular value of ε.

140 THE MONATOMIC IDEAL GAS

We will set up the required integral using the same three-dimensional quantum number space concept used previously. In this case, we wish to integrate over the total volume of one octant of this space—that is, from $R = 0$ to $R = \infty$. The number of states in a thin spherical element of this space from ε to $\varepsilon + d\varepsilon$ is given by

$$g(\varepsilon) = \frac{dG(\varepsilon)}{d\varepsilon} = \frac{d}{d\varepsilon}\left[\frac{\pi}{6}\left(\frac{8m\varepsilon}{h^2}\right)^{3/2} V\right], \tag{12.18}$$

or

$$g(\varepsilon) = \frac{\pi}{4}\left(\frac{8m}{h^2}\right)^{3/2} V\varepsilon^{1/2}. \tag{12.19}$$

In order to evaluate q_t, we must integrate the product of this $g(\varepsilon)$ and the factor $\exp(-\varepsilon/kT)$ over the full possible range of ε from zero to infinity; that is,

$$\begin{aligned}
q_t &= \int_0^\infty g(\varepsilon) e^{-\varepsilon/kT}\, d\varepsilon \\
&= \int_0^\infty \frac{\pi}{4}\left(\frac{8m}{h^2}\right)^{3/2} V\varepsilon^{1/2} e^{-\varepsilon/kT}\, d\varepsilon \\
&= \frac{\pi}{4}\left(\frac{8m}{h^2}\right)^{3/2} V \int_0^\infty \varepsilon^{1/2} e^{-\varepsilon/kT}\, d\varepsilon.
\end{aligned} \tag{12.20}$$

In performing the integration, it is convenient to introduce the dummy variable, u, defined by

$$u \equiv \frac{\varepsilon}{kT}, \qquad du \equiv \frac{d\varepsilon}{kT}. \tag{12.21}$$

Making this substitution yields

$$q_t = \frac{\pi}{4}\left(\frac{8m}{h^2}\right)^{3/2} V \int_0^\infty u^{1/2}(kT)^{1/2} e^{-u} kT\, du, \tag{12.22}$$

or

$$q_t = \frac{\pi}{4}\left(\frac{8mkT}{h^2}\right)^{3/2} V \int_0^\infty u^{1/2} e^{-u}\, du. \tag{12.23}$$

The definite integral above is the gamma function of 3/2, which is equal to $\pi^{1/2}/2$. Thus

$$q_t = \frac{\pi}{4}\left(\frac{8mkT}{h^2}\right)^{3/2} V\left(\frac{\pi^{1/2}}{2}\right), \tag{12.24}$$

or

$$q_t = \left(\frac{2\pi mkT}{h^2}\right)^{3/2} V. \tag{12.25}$$

EVALUATION OF THE PARTITION FUNCTION

It is useful now to proceed directly to the evaluation of Q in terms of q_t, neglecting q_e for the present, because in many cases the effect of electronic excitation on the calculated values of the thermodynamic functions is either small or nonexistent, and in most cases vanishes when one considers the change in a given thermodynamic property in a typical process. Making this approximation we have

$$Q = \frac{q_t^N}{N!}, \tag{12.26}$$

or

$$\ln Q = N \ln q_t - \ln N!. \tag{12.27}$$

We may make an additional approximation, namely Stirling's approximation, which states that if N is very large then

$$\ln N! \approx N \ln N - N. \tag{12.28}$$

Thus

$$\ln Q = N \ln q_t - N \ln N + N, \tag{12.29}$$

or, since we can write $N = N \ln e$, we have

$$\ln Q = N \ln q_t - N \ln N + N \ln e$$

$$= N \ln\left(\frac{q_t \cdot e}{N}\right), \tag{12.30}$$

or, by substituting for q_t we obtain

$$\ln Q = N \ln\left[\left(\frac{2\pi mkT}{h^2}\right)^{3/2}\left(\frac{Ve}{N}\right)\right]. \tag{12.31}$$

EVALUATION OF THE THERMODYNAMIC FUNCTIONS

The expressions for the various thermodynamic functions follow straightforwardly, using the previously developed relations between these functions and Q. The Helmholz free energy is

$$F = -kT \ln Q$$
$$= -NkT \ln\left[\left(\frac{2\pi mkT}{h^2}\right)^{3/2}\left(\frac{Ve}{N}\right)\right]. \tag{12.32}$$

The pressure is given by

$$P = kT\left(\frac{\partial \ln Q}{\partial V}\right)_{T,N}$$
$$= NkT \frac{\partial}{\partial V}\left(\ln\left[\left(\frac{2\pi mkT}{h^2}\right)^{3/2}\left(\frac{Ve}{N}\right)\right]\right)$$
$$= \frac{NkT}{\left(\frac{2\pi mkT}{h^2}\right)^{3/2}\left(\frac{Ve}{N}\right)} \frac{\partial}{\partial V}\left[\left(\frac{2\pi mkT}{h^2}\right)^{3/2}\left(\frac{Ve}{N}\right)\right] \tag{12.33}$$

or

$$P = \frac{NkT}{V}, \tag{12.34}$$

or

$$PV = NkT. \tag{12.35}$$

This expression has exactly the same form as the classical equation of state

$$PV = nRT, \tag{12.36}$$

leading us to the following numerical value for k:

$$k = \frac{nR}{N} = \frac{R}{N_{av}} = 1.38 \times 10^{-23} \text{ J/K}, \tag{12.37}$$

where k is known as *Boltzmann's constant*. The internal energy is given by

$$E = kT^2 \left(\frac{\partial \ln Q}{\partial T}\right)_{V,N}$$

$$= NkT^2 \frac{\partial}{\partial T}\left(\ln\left[\left(\frac{2\pi mkT}{h^2}\right)^{3/2}\left(\frac{Ve}{N}\right)\right]\right)$$

$$= \frac{NkT^2}{\left(\frac{2\pi mkT}{h^2}\right)^{3/2}\left(\frac{Ve}{N}\right)} \frac{\partial}{\partial T}\left[\left(\frac{2\pi mkT}{h^2}\right)^{3/2}\left(\frac{Ve}{N}\right)\right], \qquad (12.38)$$

or

$$E = \tfrac{3}{2}NkT. \qquad (12.39)$$

Note that the energy is a function of N and T only and is independent of V, again in agreement with the classically derived properties of the ideal gas. Note, too, that the internal energy of the system goes to zero as T goes to zero. Physically, this would correspond to having the atoms at rest in the container, separated by a distance great enough that there were no significant interatomic interactions. The magnitude of the energy, at any finite temperature, is $\tfrac{1}{2}kT$ for each of the three translational degrees of freedom of each atom. We will see that this is a general property of systems that behave classically—that is, systems in which $\Delta\varepsilon/kT$ for the mode considered is very small compared to unity.

Continuing with the evaluation of the thermodynamic functions, we may write

$$C_V = \left(\frac{\partial E}{\partial T}\right)_{N,V} = \frac{\partial}{\partial T}\left(\frac{3}{2}NkT\right) \qquad (12.40)$$

or

$$C_V = \tfrac{3}{2}Nk. \qquad (12.41)$$

The entropy is given by

$$S = \frac{E}{T} - \frac{F}{T}$$

$$= \frac{3}{2}Nk + Nk\ln\left[\left(\frac{2\pi mkT}{h^2}\right)^{3/2}\left(\frac{Ve}{N}\right)\right]$$

$$= Nk\ln\left[\left(\frac{2\pi mkT}{h^2}\right)^{3/2}\left(\frac{Ve^{5/2}}{N}\right)\right], \qquad (12.42)$$

or, after substituting $V/N = kT/P$,

$$S = Nk \ln\left[\left(\frac{2\pi mkT}{h^2}\right)^{3/2}\left(\frac{kTe^{5/2}}{P}\right)\right]. \tag{12.43}$$

This last formulation is known as the *Sackur–Tetrode equation*.

Finally, we can write

$$\mu = \frac{G}{N} = -kT\left(\frac{\partial \ln Q}{\partial N}\right)_{T,V}$$

$$= -kT \ln\left[\left(\frac{2\pi mkT}{h^2}\right)^{3/2}\left(\frac{V}{N}\right)\right], \tag{12.44}$$

or, after substituting for V/N,

$$\mu = -kT \ln\left[\left(\frac{2\pi mkT}{h^2}\right)^{3/2}\left(\frac{kT}{P}\right)\right], \tag{12.45}$$

or, after rearranging,

$$\mu = -kT \ln\left[\left(\frac{2\pi mkT}{h^2}\right)^{3/2}(kT)\right] + kT \ln P, \tag{12.46}$$

which has the same form as the classical relation

$$\mu = \mu^0(T) + kT \ln P. \tag{12.47}$$

ELECTRONIC EXCITATION

We may now return to the evaluation of the contribution to q arising from electronic excitation of the atom. This contribution can, in general, be written as

$$q_e = \sum_j \omega_{e_j} e^{-\varepsilon_{e_j}/kT}, \tag{12.48}$$

where the ω_i are the degeneracies and the ε_i the energies of the allowed electronic states of the atom, and the sum is taken over all of the j allowed electronic states. The evaluation of these terms is usually simple. The energy differences between adjacent states are in general very large. For most atoms $\Delta\varepsilon_e$ is on the order of one electron volt, or equivalent to kT at $T \approx 10^4$ K. It is thus convenient to choose as the zero of energy for this degree of freedom the electronic ground state of the atom. That is, we define

$$\varepsilon_{e_1} = 0. \tag{12.49}$$

For many atoms, in which the energy difference between the ground state and the first excited state is very large compared to kT, this is the only term in q_e that must be considered. In other cases, however, where the atom has a relatively low-lying excited state, we may find a finite fraction of the atoms in the first excited state. This is typically the case for the halogens. For fluorine, for example, $\Delta\varepsilon_{1-2} \approx 0.05$ eV. In this case, using the zero of energy defined above, we have

$$q_e = \omega_{e_1} + \omega_{e_2} e^{-\varepsilon_{e_2}/kT}, \tag{12.50}$$

where

$$\varepsilon_{e_2} = \Delta\varepsilon_{1-2},$$

the energy difference between the ground state and the first electronic excited state.

THE ZERO OF ENERGY

Recall that in the review of the laws of classical thermodynamics in Chapter 2, we noted that the first law of thermodynamics was couched in terms of the difference in energy between two different states, and that we had no basis for defining a zero of energy. Based on the above development, we are now in a position to define such a zero of energy in terms of the behavior of an ideal monatomic gas. The convention that we will adopt defines the zero of energy of the monatomic gas as the state in which the atoms are at rest in their electronic ground states, at distances far enough from one another that there are no measurable interparticle interactions. This is the reference state that we will use in all of our calculations from now on. All of the other cases that we consider will have to take account of any differences in ground state energy betwen those systems and the monatomic ideal gas ground state defined above. This will permit us to compare the state functions deduced for one system to those of other systems, as will be required in our discussions of phase and chemical equilibria. Note that this is an arbitrary definition and has no fundamental significance. It is useful, however, because it provides a reference for comparisons among different systems.

COMPLETE EXPRESSIONS FOR THE THERMODYNAMIC FUNCTIONS

We may complete our consideration of the monatomic ideal gas by writing the expressions for the thermodynamic funtions, taking into account all of the modes that contribute to q. For example, for the internal energy we may write,

$$E = NkT^2 \left[\left(\frac{\partial \ln q_t}{\partial T} \right)_{N,V} + \left(\frac{\partial \ln q_e}{\partial T} \right)_{N,V} \right]$$

$$= \frac{3}{2} NkT + \left(\frac{1}{q_e} \right) N \omega_{e_2} \varepsilon_{e_2} e^{-\varepsilon_{e_2}/kT} \qquad (12.51)$$

for the case where electronic excitation must be taken into account. For cases where electronic excitation may be neglected, the expression reverts to

$$E = \tfrac{3}{2} NkT. \qquad (12.53)$$

Evaluation of the other thermodynamic functions is straightforward. In general, one finds that, in cases where electronic excitation may be neglected, those functions that depend on a derivative of $\ln Q$ are unchanged from the values deduced without consideration of q_e. Those which depend on $\ln Q$ directly, such as S, F, G, and μ, will be changed. For example, we will have $S > S_t$, where S_t is the value of S determined from calculation using q_t only. Since typical values of ω_{e_1} are 1 to 3, the difference in entropy is on the order of $Nk \ln 2 \approx Nk$. Note, however, that although the value of S is changed, any expression for the change of S with T, P, and V will be unchanged, because the additional term in the above expression for S is independent of all of these factors.

BIBLIOGRAPHY

D. L. Goodstein, *States of Matter*, Prentice-Hall, Inc. Englewood Cliffs, NJ, 1975, Chapter 2.

T. L. Hill, *Introduction to Statistical Thermodynamics*, Addison-Wesley, 1960, Chapter 4.

J. H. Knox, *Molecular Thermodynamics*, John Wiley & Sons, New York, 1978, Chapter 6.

E. L. Knuth, *Introduction to Statistical Thermodynamics*, McGraw-Hill, New York, 1966, Chapter 8.

W. G. V. Rosser, *An Introduction to Statistical Physics*, Ellis Horwood, Chichester, UK, 1982, Chapter 5.

PROBLEMS

12.1 Develop a general expression for C_P in terms of Q. Evaluate this expression for an ideal monatomic gas.

12.2 Obtain the expression

$$\mu = -kT \ln\left[\left(\frac{2\pi mkT}{h^2}\right)^{3/2} \left(\frac{V}{N}\right)\right]$$

from the expressions

$$F = -NkT \ln\left[\left(\frac{2\pi mkT}{h^2}\right)^{3/2} \left(\frac{Ve}{N}\right)\right] \quad \text{and} \quad PV = NkT,$$

and the relation

$$N\mu = F + PV.$$

12.3 Compute the change in molal Gibbs free energy of monatomic mercury atoms (assumed to be an ideal gas) in cooling from 700 to 650 K at a constant pressure of 0.1 MPa.

12.4 Calculate the entropy per mole for argon gas at 0.1 MPa pressure at temperatures of 1000, 1 and 10^{-3} K, and extrapolate to 0 K. What does this tell you about the Sackur–Tetrode equation?

12.5 Verify the fact that the equations

$$PV = (N_1 + N_2)kT$$
$$E = \tfrac{3}{2}(N_1 + N_2)kT$$

for a binary gas mixture follow from the equation

$$Q = \frac{(q_1^{N_1} \cdot q_2^{N_2})}{(N_1! N_2!)}.$$

12.6 Derive an equation for C_{V_e}, the electronic contribution to C_V, from the equation

$$E = \frac{3}{2}kT + \left(\frac{1}{q_e}\right) N\omega_{e_2} \varepsilon_{e_2} e^{-\varepsilon_{e_2}/kT}.$$

Verify that $C_{V_e} \to 0$ as $T \to 0$ or as $T \to \infty$, and passes through a maximum in between.

12.7 Derive an expression for S_e, the electronic contribution to the entropy of a monatomic ideal gas. Check that $S_e \to k \ln(\omega_{e_1})^N$ as $T \to 0$ and that $S_e \to k \ln(\omega_{e_1} + \omega_{e_2})^N$ as $T \to \infty$.

12.8 The following spectroscopic data have been obtained for a monatomic ideal gas:

$$\omega_{e_1} = 1, \qquad \omega_{e_2} = 3, \qquad \varepsilon_{e_2} = 0.2\, eV.$$

What fraction of the atoms in the gas would be in the first excited electronic state at:
(a) 0 K
(b) 1000 K
(c) 10000 K?
(Neglect the possibility of any higher electronic excited states.)

12.9 Derive an expression for the entropy change for the following process:
Initial state: N_1 moles of ideal gas 1 at T in a container of volume V_1 plus N_2 moles of ideal gas 2 in a container of volume $V_2 = (N_2 V_1)/N_1$ at T.
Final state: N_1 moles of gas 1 and N_2 moles of gas 2 in a single container of volume $V = V_1 + V_2$ at T.

12.10 Consider a system consisting of a vessel with rigid, insulating walls, divided into two parts, each having a volume of one liter. The vessel is initially divided into two parts by a partition; this vessel contains a gas on one side, with the other side being empty. We determined in Problem 2.4 that for the process of removing the partition, $\Delta E = 0$. We also stated that $\Delta S > 0$, because this process is irreversible. Calculate ΔS for this process, using statistical thermodynamics, assuming that the gas is one mole of helium at an initial pressure of 0.1 MPa.

12.11 Using the expression for the entropy of an ideal gas given in equation (12.42), prove that the formulation $Q = q^N/N!$, rather than $Q = q^N$, is necessary in order that entropy turns out to be an extensive property of the system.

12.12 Starting with the relation for the probability that a given molecule is in a given translational energy level,

$$p_k = \frac{N(\varepsilon_k)}{N} = \frac{\Omega_k e^{-\varepsilon_k/kT}}{\sum_i e^{-\varepsilon_i/kT}},$$

where i represents the sum over states and k represents the sum over levels, and using the fact that the sum in the denominator can be replaced by an integral, as shown in equations (12.17) and (12.19), develop an expression for the molecular velocity distribution for an ideal gas. (*Hint:* Begin by determining the kinetic energy distribution.)

CHAPTER 13

THE POLYATOMIC IDEAL GAS

We may extend the treatment of the monatomic ideal gas to the case of gases made up of molecules containing more than one atom. We will do this in detail for the case of a gas of diatomic molecules, and we will discuss the general considerations involved in the treatment of molecules containing more than two atoms.

THE MODEL

While the model that we will use in this case has many similarities to that used for the monatomic gas, we will see that there is also one major difference. We will assume, as before, that the molecules behave independently of one another, and that consequently we can treat the motions of each molecule separately. However, because the two atoms of the diatomic molecule are closely bound to one another, we must also take account of the contribution to the potential energy of the system arising from the interactions between the electrons of the two atoms, which give rise to the bond energy of stable diatomic molecules. In developing the model we will also assume that we can separate this contributon to system energy arising from electronic interactions from the contributions arising from motions of the relatively massive nuclei by making use of what is called the *Born–Oppenheimer approximation*—namely, that for any given distance of separation between the two nuclei, we can consider the nuclei as fixed and study the electron motion to determine the most stable configuration consistent with the given internuclear separation. The result of this calculation gives us the contribution to the system energy due to electron motion—that is, the binding energy of the molecule.

150 THE POLYATOMIC IDEAL GAS

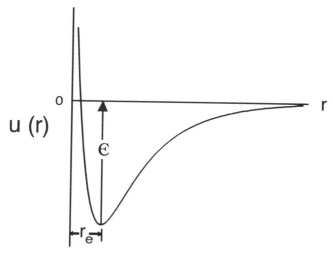

Figure 13.1 The potential energy curve for a system consisting of two atoms, in which the potential energy versus internuclear separation is governed by the Lennard-Jones 6-12 potential.

The next question is that of the measurement or calculation of this interaction potential. A number of empirical mathematical forms for the interaction potential have been developed over the years to describe observed experimental results. All of these have the property that the system potential energy, referenced to the zero of energy defined in Chapter 12, at first decreases as the two atoms are brought together, then passes through a minimum, followed by a sharp rise as the atoms are brought very close together. This behavior is most commonly modeled by an equation of the form

$$u(r) = -\frac{A}{r^n} + \frac{B}{r^m}, \qquad (13.1)$$

in which n and m are integers. Such a potential is often referred to as a *pair potential*, and will be seen again when we treat condensed phase systems in Chapters 14 through 16. A typical result of a pair potential calculation is shown schematically in Figure 13.1. Here we plot the potential energy of a molecule as a function of the internuclear distance, r, for the commonly used case where $n = 6$, $m = 12$, $A = 2\varepsilon r_e^6$, and $B = \varepsilon r_e^{12}$ (the so-called Lennard-Jones potential). Note that, consistent with the zero of energy defined in Chapter 12, we have $u(r) = 0$ at $r = \infty$ and that the shape of the well is defined by the maximum well depth, ε (a positive number), and the equilibrium interatomic spacing, r_e. Values of ε and r_e for a number of common gases are given in Appendix E. Note, too, that this is just one of many possible expressions for the form of the potential well. The appropriate form in any case will depend

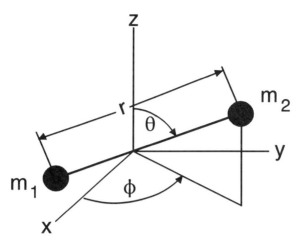

Figure 13.2 An illustration of the parameters that define the center-of-mass coordinate system for a diatomic molecule.

on the details of the electronic interactions for that case. In the case of diatomic molecules, the maximum well depth is commonly called D_e, the dissociation energy of the molecule. This is the energy that accounts for the contribution of the electron motion to the energy of the system, and it will give rise to one term in the system partition function. Note that as we have defined it, D_e is is inherently positive.

Consider now the energy associated with the motion of the nuclei. To treat this motion quantum mechanically, we could write the Schrödinger equation for each molecule in terms of the positions of each of the two nuclei—for example, x_1, y_1, z_1; x_2, y_2, z_2, the coordinates of the two atoms in a box of volume V—similar to the procedure used for the monatomic ideal gas. As a practical matter, since the two atoms are bound together, it is more convenient to consider their motion as a unit in a so-called center-of-mass coordinate system. The relationship between the two coordinate systems is shown in Figure 13.2. The relevant parameters in the center-of-mass coordinate system are the x, y, z position of the center of mass of the molecule, the internuclear separation, r, and the orientation of the axis of the molecule, defined by the angles, Φ and θ.

Note that in both coordinate systems, six coordinates (three per atom) must be specified in order to specify the position of the molecule. We may describe the modes of motion available to the molecule, in terms of the center-of-mass coordinate system, as translation in three orthogonal directions of the center of mass of the molecule, motion of the two nuclei relative to one another along r, which is essentially a vibrational motion, and rotation of the molecule about the axes defined by Φ and θ.

152 THE POLYATOMIC IDEAL GAS

We will assume that the vibration can be described as that of a one-dimensional harmonic quantum oscillator and that the rotations can be described as those of a rigid quantum rotator. These assumptions imply that all of the molecular modes of motion are independent of one another, and that we can write the total energy of any molecule as

$$\varepsilon = \varepsilon_t + \varepsilon_v + \varepsilon_r + \varepsilon_e, \tag{13.2}$$

where the subscripts t, v, r, and e refer to translational, vibrational, rotational, and electronic contributions, respectively. Since the molecules are independent, identical, and indistinguishable, and since we have already shown that the number of available translational energy states is very large compared to the number of particles in a typical system, we may use the expression

$$Q = \frac{q^N}{N!}, \tag{13.3}$$

where in this case

$$q = q_t \cdot q_v \cdot q_r \cdot q_e, \tag{13.4}$$

to calculate the partition function for this system.

EVALUATION OF q

Let us now go through the process of evaluating the contributions to q of the various q_i listed above. The translational modes are treated identically to the previously discussed case of the monatomic ideal gas. The only difference in the present case is that the mass of the translating species is $(m_1 + m_2)$, the sum of the masses of the two atoms that make up the diatomic molecule, rather than the mass of the single atom of the previous case. (Recall that in the previous development, no mention was made of the structure of the particle; the only property of significance was its total mass.) We may thus write, for the diatomic molecule, that

$$q_t = \left[\frac{2\pi(m_1 + m_2)kT}{h^2}\right]^{3/2} V. \tag{13.5}$$

The electronic contribution to the partition function, associated with the fact that the potential energy of the bound system of two atoms is less than the energy of the two atoms at large separation, can be written, in general, as

$$q_e = \sum_i \omega_{e_i} e^{-\varepsilon_{e_i}/kT}, \tag{13.6}$$

where the ε_{e_i} are the energies of the allowed electronic states of the molecule. In most cases, we will have to consider only the ground electronic state of the molecule. This is because the energy difference between electronic states in diatomic molecules is typically greater than one electron volt and is consequently very large compared to kT for most cases we will consider. If this is the case, then, in order to be consistent with our previously chosen zero of energy, we will have

$$\varepsilon_{e_1} = -D_e, \tag{13.7}$$

and consequently

$$q_e = \omega_{e_1} e^{D_e/kT}. \tag{13.8}$$

In the event that electronic excited states of the molecule must be taken into consideration, one must, of course, include terms in the partition function to account for this excitation. For example, if the first excited state has a significant probability of occupation we must use

$$q_e = \omega_{e_1} e^{D_e/kT} + \omega_{e_2} e^{-\varepsilon_{e_2}/kT}, \tag{13.9}$$

where ε_{e_2} is the energy of the molecular excited state relative to the zero of energy based on the separated atoms in their electronic ground states and is given by

$$\varepsilon_{e_2} = -D_e + \Delta\varepsilon_{1-2}, \tag{13.10}$$

in which $\Delta\varepsilon_{1-2}$ is the energy difference between the electronic ground state and the first electronic excited state.

We may treat the vibrational motion of the molecule by assuming that the potential well associated with the binding energy of the molecule is essentially parabolic near the bottom. In this case, $u(r)$ near the bottom of the well will be given by

$$u(r) = -D_e + \tfrac{1}{2}f(r - r_e)^2, \tag{13.11}$$

where the force constant, f, is

$$f = \left(\frac{\partial^2 u(r)}{\partial r^2}\right)_{r_e} \tag{13.12}$$

and we have neglected fourth and higher-order terms. The equation of motion for vibrational motion of the diatomic molecule in such a well is

$$\mu\ddot{x} = -fx, \tag{13.13}$$

154 THE POLYATOMIC IDEAL GAS

in which

$$x = (r - r_e) \quad \text{and} \quad \mu = \frac{m_1 m_2}{m_1 + m_2},$$

where μ is the so-called reduced mass of the molecule. Solution of this equation yields

$$v = \left(\frac{1}{2\pi}\right)\left(\frac{f}{\mu}\right)^{1/2} \tag{13.14}$$

as the appropriate vibrational frequency. The energy levels available to a one-dimensional quantum harmonic oscillator, as developed in our treatment of quantum mechanics in Chapter 10, are given by

$$\varepsilon_n = (n + \tfrac{1}{2})hv, \quad n = 0, 1, 2, \ldots. \tag{13.15}$$

Consequently,

$$q_v = \sum_{n=0}^{\infty} e^{-\varepsilon_n/kT} = e^{-hv/2kT} \sum_{n=0}^{\infty} e^{-nhv/kT}. \tag{13.16}$$

The last sum on the right-hand side of the equation converges and has the value

$$\sum_{n=0}^{\infty} e^{-nhv/kT} = \frac{1}{1 - e^{-hv/kT}}. \tag{13.17}$$

Thus

$$q_v = \frac{e^{-hv/2kT}}{1 - e^{-hv/kT}} \tag{13.18}$$

or

$$q_v = \frac{e^{-\Theta_v/2T}}{1 - e^{-\Theta_v/T}}, \tag{13.19}$$

where we have made the substitution

$$\Theta_v \equiv \frac{hv}{k}, \tag{13.20}$$

and Θ_v is called the *vibrational temperature of the molecule*. Typical values of Θ_v are on the order of 3000 K. Consequently, at temperatures around room temperature, very few molecules will be in excited vibrational states. Significant vibrational excitation will be found at room temperature only for a few molecules having high masses and relatively weak binding. For other molecules, very high temperatures are required for significant vibrational excitation.

We may finally consider the contribution of the rotation of the molecule as a whole about two independent axes of rotation. We will treat the molecule as a rigid quantum rotator. The allowed energy levels of such a system are given, again recalling our discussion of quantum mechanics, as

$$\varepsilon_j = \frac{j(j+1)h^2}{8\pi^2 I}, \qquad j = 0, 1, 2, \ldots, \qquad (13.21)$$

where I is the moment of inertia of the molecule. There is an additional complication in this case, since the rotational energy levels are inherently degenerate, with a degeneracy given by

$$\omega_j = 2j + 1. \qquad (13.22)$$

Consequently, the expression for q_r can be written

$$q_r = \sum_{j=0}^{\infty} \omega_j e^{-\varepsilon_j/kT}$$

$$= \sum_{j=0}^{\infty} (2j+1) e^{-j(j+1)h^2/8\pi^2 IkT}, \qquad (13.23)$$

or

$$q_r = \sum_{j=0}^{\infty} (2j+1) e^{-j(j+1)(\Theta_r/T)}, \qquad (13.24)$$

where we have defined the rotational temperature, Θ_r, as

$$\Theta_r \equiv \frac{h^2}{8\pi^2 Ik}, \qquad (13.25)$$

analogous to the previously defined Θ_v. Typical values of Θ_r, however, are much smaller than the typical values of Θ_v. Generally, Θ_r is a few degrees Kelvin. Consequently, except for very light molecules at very low temperatures, the spacing between rotational energy levels will be small compared to kT. As we have seen previously for the case of translational degrees of freedom, this allows us to replace the sum in the expression for q_r by an integral. Thus, except at very low temperature, we may write

$$q_r = \int_0^\infty (2j + 1)e^{-j(j+1)(\Theta_r/T)} \, dj, \qquad (13.26)$$

or

$$q_r = \int_0^\infty e^{-j(j+1)(\Theta_r/T)} \, d[j(j+1)], \qquad (13.27)$$

giving

$$q_r = \frac{T}{\Theta_r} \qquad (13.28)$$

as the expression for q_r.

There is, however, one final complication. The above derivation has assumed implicitly that the masses of the two atoms making up the diatomic molecule were different. In the event that we are considering a homonuclear diatomic molecule—that is, one with the two atoms identical—the above relation overcounts the number of available rotational states by a factor of two. This overcounting may be taken care of by dividing the final expression by a term, σ, called the *symmetry number*. For a homonuclear (or symmetrical) diatomic molecule, $\sigma = 2$; otherwise $\sigma = 1$. We thus have, finally,

$$q_r = \frac{T}{\sigma \Theta_r}. \qquad (13.29)$$

EVALUATION OF THE THERMODYNAMIC FUNCTIONS

We now have a complete expression for q for the diatomic molecule, and through the relations between q and Q and the relations relating the thermodynamic functions to Q, we have all the information we need to evaluate the thermodynamic functions. We may write the expression for Q as

$$Q = \frac{q^N}{N!} = \frac{1}{N!} (q_t \cdot q_v \cdot q_r \cdot q_e)^N, \qquad (13.30)$$

or

$$\ln Q = -N \ln N + N + N \ln q_t + N \ln q_v + N \ln q_r + N \ln q_e, \qquad (13.31)$$

or

$$\ln Q = N\left(\ln \frac{q_t e}{N} + \ln q_r + \ln q_v + \ln q_e\right). \quad (13.32)$$

As usual, the expressions for the thermodynamic functions follow straightforwardly. For the case where we may neglect electronic excited states of the molecule, we have

$$F = -NkT \ln\left[\left(\frac{2\pi(m_1 + m_2)kT}{h^2}\right)^{3/2}\left(\frac{Ve}{N}\right)\right] - NkT \ln\left(\frac{8\pi^2 IkT}{\sigma h^2}\right)$$
$$+ \left(\frac{Nh\nu}{2}\right) + NkT \ln(1 - e^{-h\nu/kT}) - ND_e + NkT \ln \omega_{e_1}. \quad (13.33)$$

$$E = \frac{3}{2}NkT + \frac{2}{2}NkT + \frac{Nh\nu}{2} + \frac{Nh\nu}{e^{h\nu/kT} - 1} - ND_e. \quad (13.34)$$

$$C_V = \frac{3}{2}Nk + \frac{2}{2}Nk + Nk\left(\frac{h\nu}{kT}\right)^2 \frac{e^{h\nu/kT}}{(e^{h\nu/kT} - 1)^2} \quad (13.35)$$

$$S = Nk \ln\left[\left(\frac{2\pi(m_1 + m_2)kT}{h^2}\right)^{3/2}\left(\frac{Ve^{5/2}}{N}\right)\right] + Nk \ln\left(\frac{8\pi^2 IkTe}{\sigma h^2}\right)$$
$$+ Nk\left(\frac{\frac{h\nu}{kT}}{[e^{h\nu/kT} - 1]}\right) - Nk \ln(1 - e^{-h\nu/kT}) + Nk \ln \omega_{e_1}, \quad (13.36)$$

$$\mu = -kT \ln\left[\left(\frac{2\pi(m_1 + m_2)kT}{h^2}\right)^{3/2}\frac{V}{N}\right] - kT \ln\left(\frac{8\pi^2 IkT}{\sigma h^2}\right)$$
$$+ \frac{h\nu}{2} + kT \ln(1 - e^{-h\nu/kT}) - D_e - kT \ln \omega_{e_1}. \quad (13.37)$$

and finally

$$PV = NkT, \quad (13.38)$$

as was the case for the monatomic ideal gas. Thus all that we need to evaluate these thermodynamic functions from first principles is a knowledge of (a) the masses of the atoms involved, (b) the factors that control Θ_v and Θ_r, namely ν and I, and (c) the dissociation energy of the molecule. These parameters can be obtained experimentally from the optical spectra of the molecule. The appropriate Θ_v, Θ_r, and D_e data for a number of diatomic molecules are presented in Appendix F.

POLYATOMIC MOLECULES

Let us next consider briefly the complications that arise for a gas of molecules containing more than two atoms per molecule—that is, polyatomic molecules. The general rule that the number of degrees of motional freedom is three per atom in the molecule that we observed in the case of the diatomic molecule holds as well for larger molecules. In all cases, the molecule has three translational degrees of freedom. The remaining degrees of freedom are divided among vibrational and rotational modes. The number of available rotational modes is always either two or three. Linear molecules, such as CO_2, have two rotational modes. Nonlinear molecules have three rotational modes, which may have different values of Θ_r, associated with the different values of I for the three rotational axes. The remaining available modes, $(3n - 5)$ in the case of a linear polyatomic molecule and $(3n - 6)$ in the case of a nonlinear molecule, will be vibrational modes. Each of these modes will be represented in the partition function by a term of the type given in equation (13.19), with the appropriate value of v. Information on the moments of inertia and the vibrational frequencies for any given molecule are obtained from optical spectroscopic measurements, as in the case of diatomic molecules. Rotational and vibrational data for some common polyatomic molecules are given in Appendix F.

THE GRAND CANONICAL ENSEMBLE

As a final point before we leave the subject of ideal gases, let us return briefly to the question of which ensemble is appropriate in a given case. In our treatment of ideal gases we have, up to this point, used the canonical ensemble. This was a useful choice, because the expressions for the translational energy levels were stated explicitly in terms of the volume of the system. Let us now look at the treatment of the ideal gas in terms of the grand canonical ensemble. We defined the grand canonical ensemble partition function as

$$\Xi = \sum_{i,N} e^{-E_i(N)/kT} e^{-N\mu/kT}, \qquad (13.39)$$

which can also be written as

$$\Xi = \sum_N Q(NVT) e^{-N\mu/kT}. \qquad (13.40)$$

We also determined that

$$PV = kT \ln \Xi, \qquad (13.41)$$

or, alternatively,

$$\ln \Xi = \frac{PV}{kT}. \tag{13.42}$$

If we apply the grand canonical ensemble to a system of independent, identical, indistinguishable particles, for which the number of available energy states is large compared to the number of particles, we may substitute

$$Q(NVT) = \frac{q^N}{N!}, \tag{13.43}$$

leading to

$$\Xi = \sum_N \left(\frac{q^N}{N!}\right) e^{-N\mu/kT}. \tag{13.44}$$

The value of this sum can be shown to equal

$$\Xi = e^{qe^{\mu/kT}}, \tag{13.45}$$

or

$$\ln \Xi = qe^{\mu/kT}. \tag{13.46}$$

When we determined the relation between Ξ and PV, we also deduced that

$$N = kT\left(\frac{\partial \ln \Xi}{\partial \mu}\right)_{V,T}, \tag{13.47}$$

or, for this case

$$N = kT\frac{\partial}{\partial \mu}(qe^{\mu/kT})_{V,T}. \tag{13.48}$$

Since $q = f(V, T)$ only, evaluation of the derivtive yields

$$N = kT\left(\frac{1}{kT}\right) qe^{\mu/kT}, \tag{13.49}$$

or

$$N = qe^{\mu/kT} = \ln \Xi. \tag{13.50}$$

Finally, since

$$\ln \Xi = \frac{PV}{kT} \qquad (13.51)$$

we have

$$N = \frac{PV}{kT}, \qquad (13.52)$$

or

$$PV = NkT, \qquad (13.53)$$

as we deduced using the canonical ensemble. Thus we see that the ideal gas equation of state is a general equation of state for any system of independent, identical, indistinguishable particles, for the case where the number of available states is large compared to the number of particles. Our treatment of the ideal monatomic and diatomic gases, using the canonical ensemble, showed this to be the case for those two systems. We now know that it is valid for *all* ideal gases, independent of the nature of the molecules involved. On the other hand, when we consider the electron gas in Chapter 25, we will see that this equation of state does *not* apply to the case where the number of available states is *not* large compared to the number of particles.

BIBLIOGRAPHY

T. L. Hill, *Introduction to Statistical Thermodynamics*, Addison-Wesley, 1960, Chapters 8 and 9, Appendix D.

J. H. Knox, *Molecular Thermodynamics*, John Wiley & Sons, New York, 1978, Chapter 6 and 9.

E. L. Knuth, *Introduction to Statistical Thermodynamics*, McGraw-Hill, New York, 1966, Chapter 13.

W. G. V. Rosser, *An Introduction to Statistical Physics*, Ellis Horwood, Chichester, UK, 1982, Chapter 5.

J. W. Whalen, *Molecular Thermodynamics*, John Wiley & Sons, New York, 1991, Chapter 5, Appendix D.

PROBLEMS

13.1 Develop an expression for C_P of an ideal diatomic gas in terms of observables and evaluate it for nitrogen gas at 300 K and 1000 K.

13.2 One mole of $^{16}O^{16}O$ and one mole of $^{18}O^{18}O$ are mixed together in a thermostatted one-liter box at 300 K. Initially, no reaction takes place.

Later, a catalyst is added which catalyzes the exchange reaction

$$(^{16}O^{16}O) + (^{18}O^{18}O) \rightleftharpoons 2(^{16}O^{18}O)$$

and this reaction proceeds to equilibrium.
(a) What is the energy change for this reaction?
(b) What is the entropy change for this reaction?
(c) How would one determine the equilibrium gas composition?

13.3 Derive an equation for the contribution to the entropy of an ideal diatomic gas that results from its rotational energy.

13.4 Calculate the entropy change for the following process: One mole of chlorine gas, initially at 200 K and 0.1 MPa, is heated isobarically to 800 K. You may assume that chlorine behaves as an ideal gas in this range of temperature and pressure.

13.5 Calculate the heat flow and the work done when one mole of bromine is cooled reversibly and isobarically from 1000 K to 500 K at a pressure of 0.1 MPa. You may neglect the possibility of dissociation of the diatomic molecule.

13.6 In the CO_2 molecule, the two oxygen atoms are bonded to the carbon atom. Describe a simple thermodynamic measurement that would tell you whether the structure corresponds to Figure A or Figure B below.

$$O=C=O \qquad O=C\overset{\displaystyle O}{\|}$$
(A) (B)

13.7 The following data are available for hydrogen (H_2) and deuterium (D_2):

H_2	D_2
$D_e = 4.45$ eV	$D_e = 4.45$ eV
$r_e = 0.741$ Å	$r_e = 0.741$ Å
$\omega_{e_1} = 1$	$\omega_{e_1} = 1$
$M_H = 1$ amu	$M_D = 2$ amu
$\Theta_v = 6210$ K	
$\Theta_r = 85.4$ K	

(a) What would be the difference in internal energy between one mole of H_2 and one mole of D_2 at 500 K?
(b) Qualitatively, what terms in the expression for the internal energy would be treated differently at 50 K? at 10,000 K?
(Hint: How do Θ_v and Θ_r depend on M?)

13.8 Calculate the difference in C_V per mole for H_2 and D_2 at 2000 K. You may assume that both species behave as ideal diatomic gases, and you may neglect dissociation and electronic excitation of the molecules.

13.9 What is the change in Gibbs free energy when one mole of N_2 molecules in a 10-liter container is heated from 300 K to 1500 K?. Assume that no dissociation takes place.

CHAPTER 14

THE EINSTEIN MODEL OF THE SOLID

In this chapter and the following chapter, we will consider models of crystalline solids. The critical requirement of any such model is to characterize the energy associated with the vibrations of the atoms about their equilibrium sites in the crystal. In this chapter, we will consider the mathematically simple Einstein model. In the next chapter, we will see how the use of a more sophisticated model, the Debye model, both improves the description of the behavior of the system and increases the mathematical complexity of the treatment.

THE EINSTEIN MODEL

The model in this case considers the crystal to consist of identical atoms, each bound to a lattice point of a three-dimensional lattice of fixed sites, each vibrating about its equilibrium position with a vibrational amplitude that is small enough that no place exchange is possible. Each atom is assumed to vibrate independently of its neighbors and to execute a harmonic vibration in three dimensions in the assumedly spherical potential well formed by the interaction of the atom with its neighbors.

This choice of model dictates the most convenient ensemble to use for the statistical mechanical treatment of the system. Since the atoms are fixed to sites, they are, in principle, distinguishable. Since there is a fixed number of atoms per unit volume of crystal, an ensemble of fixed N, V, and T—that is, a canonical ensemble—is the obvious choice. What we must treat, then, is a system of independent, identical, distinguishable particles, each vibrating in a spherically symmetric potential well. We will assume that the only effect that the presence of one atom has on another is to contribute to the formation of the potential well, and that the vibrational state of one atom is independent of the vibrational states of all the other atoms.

In evaluating the partition function for this system, we must consider two contributions to the energy of each microstate, namely, the potential energy associated with the interactions of the atoms with one another, which gives rise to the potential well mentioned above, and the energy associated with the vibrational motion of the atoms within the potential wells.

Consider first the energy associated with the potential wells. We may treat this by extension of the treatment of the binding energy of the diatomic molecule given in Chapter 13, because it is essentially the energy difference between N atoms at rest, far apart, in their atomic ground state that we chose as our zero of energy, and N atoms bound to lattice sites separated by an equilibrium distance r_0. A common way of treating this problem is to assume that each atom in the system interacts with each other atom in the system according to a potential energy versus distance curve similar to that described in Chapter 13, and that the total interaction energy is just the sum of these so-called pairwise interactions, as shown in Figure 14.1 for a simple one-dimensional case. The resulting potential energy for an N-atom system is

$$U(N) = \frac{1}{2} \sum_{i \neq j} u(r_{ij}), \tag{14.1}$$

where $u(r_{ij})$ is the value of the pairwise interaction potential for an interatomic distance r_{ij}. The sum is taken over all of the atoms in the crystal, and the factor

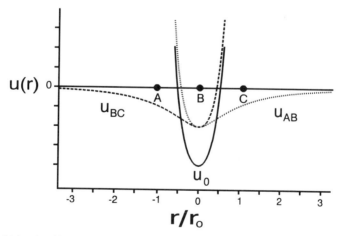

Figure 14.1 An illustration of the potential well created about atom B due to the superposition of the two pairwise interactions arising from its interaction with atoms A and C in a linear chain. In a three-dimensional system, the resulting potential well would have spherical symmetry for small departures from u_0.

$\frac{1}{2}$ is included to avoid overcounting [i.e., the interaction between any two atoms i and j should not be counted both as $u(r_{ij})$ and $u(r_{ji})$]. The potential well depth in this case of a crystalline solid will be given by

$$u_0 = \sum_r u(r)n(r), \qquad (14.2)$$

where the $n(r)$ are the number of atoms at each given equilibrium distance from the reference atom. It can be seen that for a typical case, since $n(r)$ increases roughly as r^3, and $u(r)$ decreases roughly as r^{-6}, the contribution from atoms at large distances will be small, decreasing roughly as r^3. Note that, as in the previous case, u_0 will be inherently negative.

Because of the rapid decrease in the contribution to u_0 from atoms at increasing distances, it is acceptable in many cases to include in the sum of equation (14.2) only terms from a limited number of neighbors. Note that, for a given pair potential, the value of u_0 will depend on the crystal structure, and that, all other things being equal, close packed structures will be favored.

In the extreme case, it is possible to describe the interaction energy in terms of nearest neighbors only, the so-called nearest-neighbor model. In this case, which we will see again when we discuss multicomponent condensed phases, we may write

$$u_0 = cu(r_0), \qquad (14.3)$$

and from this

$$U(N) = \frac{Nu_0}{2} = \frac{Ncu(r_0)}{2}, \qquad (14.4)$$

in which c is the coordination number appropriate to the crystal structure of the material and $u(r_0)$ is the value of the interaction potential at the equilibrium interatomic distance in the solid. In any case, the term $Nu_0/2$ represents the total binding energy of the system, and $N_{av}u_0/2$ is equivalent to $-(\Delta E_s)_0$, the molar energy of sublimation at 0 K.

Consider next the vibrational motion of a single atom. The three-dimensional vibration can be broken up into contributions along the three principal axes of the crystal, x, y, and z. Subject to the assumptions made above, the potential energy of the atom in its potential well will be

$$u(x, y, z) = u_0 + \left(\frac{f}{2}\right)(x^2 + y^2 + z^2), \qquad (14.5)$$

where u_0 again is the depth of the potential well at the lattice point; x, y, and z represent the instantaneous displacement of the atom from this lattice point;

166 THE EINSTEIN MODEL OF THE SOLID

and f, the force constant for the vibrational motion, is given by

$$f = \left(\frac{\partial^2 u}{\partial x^2}\right)_0 = \left(\frac{\partial^2 u}{\partial y^2}\right)_0 = \left(\frac{\partial^2 u}{\partial z^2}\right)_0. \quad (14.6)$$

That is, f depends on the shape of the potential well, which in turn will depend on the volume per atom in the crystal and the binding energy of the crystal. We have neglected fourth-order and higher-order contributions to f.

For this form of u, all three vibrational modes of a given atom are equivalent. This, coupled with the previous assumptions that all atoms are identical and independent, means that the crystal as a whole has 3N identical, independent vibrational modes. Note here that the general rule that there are three degrees of motional freedom per atom, which we first discussed in connection with the diatomic molecule, is also obeyed in this case. What we see here is that the three degrees of translational freedom that each atom has in the monatomic ideal gas appear as three degrees of vibrational freedom in the crystal.

EVALUATION OF THE PARTITION FUNCTION

We may now write an expression for the energy of any microstate of the macroscopic crystal as

$$E_j = \frac{Nu_0}{2} + \sum_{i=1}^{3N} \varepsilon_{n_i}, \quad (14.7)$$

where the term $Nu_0/2$ accounts for the binding energy of the crystal and the second term is the sum, over all of the $3N$ vibrational modes, of the energy of each mode associated with the energy state E_j of the macroscopic system.

We may now write an expression for the partition function of this system. Since we are applying the canonical ensemble to a system of independent, identical, distinguishable particles, the appropriate form for Q is

$$Q = q^N. \quad (14.8)$$

We may further expand this expression, separating the contributions arising from the binding energy from the contributions arising from the $3N$ vibrational modes to yield

$$Q = q_0^N \cdot q_v^{3N}, \quad (14.9)$$

in which

$$q_0 = e^{-u_0/2kT}. \quad (14.10)$$

Thus

$$Q = e^{-Nu_0/2kT} \cdot q_v^{3N}. \tag{14.11}$$

We may now proceed with the evaluation of q_v. The previously made assumptions concerning the shape of the potential well and the magnitude of the vibrational motion dictate that the motion is essentially harmonic. We can write the x-component of the potential energy of a moving atom, relative to the potential energy of that atom at rest at the bottom of the potential well, as

$$u(x) = u_x - u_0 = \frac{fx^2}{2}. \tag{14.12}$$

The vibrational frequency associated with this motion in a well of this shape is given by classical mechanics as

$$v = \frac{1}{2\pi}\left(\frac{f}{m}\right)^{1/2}, \tag{14.13}$$

where m is the mass of the atom. The quantum mechanically allowed values of the energy levels of such an oscillator are, as shown previously,

$$\varepsilon_n = (n + \tfrac{1}{2})hv, \quad n = 0, 1, 2, \ldots. \tag{14.14}$$

Thus

$$q_v = \sum_{n=0}^{\infty} e^{-(n+1/2)hv/kT} = e^{-hv/2kT} \sum_{n=0}^{\infty} e^{-nhv/kT}. \tag{14.15}$$

In general, the calculated value of v for real crystals is on the order of 10^{12} sec^{-1}. At ordinary temperatures, this gives a value of hv/kT of order unity. Consequently, the terms $\exp(-nhv/kT)$ are small compared to unity. For this case, the series given above converges and has the value

$$\sum_{n=0}^{\infty} e^{-nhv/kT} = \frac{1}{1 - e^{-hv/kT}}. \tag{14.16}$$

Thus

$$Q = e^{-Nu_0/2kT} \left[\frac{e^{-hv/2kT}}{1 - e^{-hv/kT}}\right]^{3N} \tag{14.17}$$

is the final expression for the partition function of the system.

168 THE EINSTEIN MODEL OF THE SOLID

It is convenient to rewrite this expression in terms of a parameter called the *Einstein temperature*, defined by

$$\Theta_E \equiv \frac{h\nu}{k}, \tag{14.18}$$

yielding

$$Q = e^{-Nu_0/2kT} \left[\frac{e^{-\Theta_E/2T}}{1 - e^{-\Theta_E/T}} \right]^{3N}. \tag{14.19}$$

We may also arrive at another useful form for Q by making the substitution

$$\frac{e^{-\Theta_E/2T}}{1 - e^{-\Theta_E/T}} = \frac{1}{e^{\Theta_E/2T} - e^{-\Theta_E/2T}} = \frac{1}{2\sinh\left(\frac{\Theta_E}{2T}\right)}, \tag{14.20}$$

yielding

$$Q = e^{-Nu_0/2kT} \left[\frac{1}{2\sinh\left(\frac{\Theta_E}{2T}\right)} \right]^{3N}. \tag{14.21}$$

We will use both of these last two formulations for Q in developing and examining the relations for the thermodynamic functions of this sytem. For convenience in calculations involving the Einstein treatment, a table of hyperbolic functions is given in Appendix G.

LIMITING VALUES OF q_v

Before we develop explicit expressions for the thermodynamic functions, let us look at the limiting values of q_v at high and low temperatures, because we will use these limits later. At low temperatures—that is, T small compared to Θ_E—we have

$$q_v = \frac{e^{-\Theta_E/2T}}{1 - e^{-\Theta_E/T}} \approx e^{-\Theta_E/2T}. \tag{14.22}$$

In other words, all of the vibrational modes will be in the ground vibrational state. This implies that the atoms in the crystal will never be at rest. Even at very low temperatures there will be this ground-state vibration. For this case

$$Q \to e^{-N(u_0/2kT + 3\Theta_E/2T)}. \tag{14.23}$$

At high temperature—that is, T large compared to Θ_E—we have

$$q_v = \frac{1}{2\sinh\left(\dfrac{\Theta_E}{2T}\right)} = \frac{T}{\Theta_E}, \tag{14.24}$$

as $\sinh x \approx x$ at small values of x. This expression may be compared to the expression for q_r in the treatment of the diatomic molecule. Again in this case, if the separation between energy levels is small compared to kT, essentially classical behavior is observed. The above form for q_v leads to

$$Q = e^{-Nu_0/2kT}\left(\frac{T}{\Theta_E}\right)^{3N}. \tag{14.25}$$

EVALUATION OF THE THERMODYNAMIC FUNCTIONS

We may now proceed to write expressions for the thermodynamic functions in terms of Q for the Einstein solid, using the general form of

$$\ln Q = -\frac{Nu_0}{2kT} - 3N \ln\left[2\sinh\left(\frac{\Theta_E}{2T}\right)\right]. \tag{14.26}$$

The Helmholz free energy is

$$F = -kT \ln Q$$
$$= \frac{Nu_0}{2} + 3NkT \ln\left[2\sinh\left(\frac{\Theta_E}{2T}\right)\right]. \tag{14.27}$$

The internal energy is

$$E = kT^2 \left(\frac{\partial \ln Q}{\partial T}\right)_{N,V}$$
$$= kT^2 \left[\frac{Nu_0}{2kT^2} - 3N \frac{\partial}{\partial T}\left(\ln 2\sinh\left(\frac{\Theta_E}{2T}\right)\right)\right]$$
$$= \frac{Nu_0}{2} + \frac{3Nk\Theta_E}{2} \coth\left(\frac{\Theta_E}{2T}\right). \tag{14.28}$$

The entropy is given by

$$S = \frac{E}{T} - \frac{F}{T}$$

$$= \frac{Nu_0}{2T} + 3Nk\left(\frac{\Theta_E}{2T}\right)\coth\left(\frac{\Theta_E}{2T}\right) - \frac{Nu_0}{2T} - 3Nk\ln\left[2\sinh\left(\frac{\Theta_E}{2T}\right)\right]$$

$$= 3Nk\left[\left(\frac{\Theta_E}{2T}\right)\coth\left(\frac{\Theta_E}{2T}\right) - \ln\left(2\sinh\left(\frac{\Theta_E}{2T}\right)\right)\right]. \tag{14.29}$$

The heat capacity, C_V, is

$$C_V = \left(\frac{\partial E}{\partial T}\right)_{N,V}$$

$$= \frac{\partial}{\partial T}\left[\frac{Nu_0}{2} + \frac{3}{2}Nk\Theta_E\coth\left(\frac{\Theta_E}{2T}\right)\right]$$

$$= 3Nk\left(\frac{\Theta_E}{2T}\right)^2\left[\frac{1}{\sinh\left(\frac{\Theta_E}{2T}\right)}\right]^2. \tag{14.30}$$

Finally, since we have $PV \ll E$ for a crystalline solid, we may say, approximately, that

$$H \approx E, \tag{14.31}$$

$$G = N\mu \approx F, \tag{14.32}$$

and finally

$$\mu = \frac{u_0}{2} + 3kT\ln\left[2\sinh\left(\frac{\Theta_E}{2T}\right)\right]. \tag{14.33}$$

The forms of the expressions for E, S, and C_V may be shown graphically by using what is known as a "reduced temperature" plot, where the thermodynamic functions, in appropriate form, are plotted versus T/Θ_E. It can be seen from the expressions for E, S, and C_V that they depend only on N, u_0, and T/Θ_E. Thus if one plots $[E - (Nu_0/2)]/(Nk\Theta_E)$ or S/Nk or C_V/Nk versus T/Θ_E, one should obtain a universal plot for all materials. That is, the experimentally measured values for these parameters for all systems that are described by the Einstein model should appear on the same set of curves, when a proper choice of Θ_E is made. The appropriate universal curves are shown in Figure 14.2.

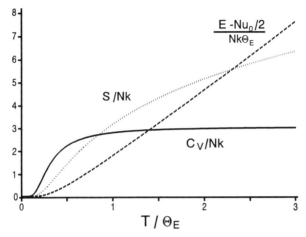

Figure 14.2 A reduced temperature plot showing the temperature variation of the thermodynamic properties of crystalline solids according to the Einstein model in terms of the reduced temperature, T/Θ_E.

As a practical matter, the curve used as a criterion for how well the theory fits experimental observation is the curve of C_V/Nk versus T/Θ_E. Typically, one finds that with proper choice of Θ_E, a good fit can be obtained for many materials at moderate-to-high temperatures. Deviations from the Einstein relation are generally observed at low temperatures. The experimental points lie increasingly above the theoretical curve as the temperature approaches zero. This discrepancy is connected with the fact that the model assumes that only one vibrational frequency, and a fairly high one at that, is available to the crystal. The low probability of excitation of this high-frequency mode at low temperature leads to a systematic underestimate of C_V at low temperatures.

HIGH AND LOW TEMPERATURE LIMITS

Let us finish our discussion of the Einstein model by looking at the predicted behavior of the thermodynamic functions at very low and very high temperatures. At low temperatuares, T small compared to Θ_E, we have

$$F \to \frac{Nu_0}{2} + 3NkT\left(\frac{\Theta_E}{2T}\right)$$

$$\to \frac{Nu_0}{2} + \frac{3Nk\Theta_E}{2}. \tag{14.34}$$

That is, F approaches a constant value independent of temperature. Moreover,

$$E \to \frac{Nu_0}{2} + \frac{3Nk\Theta_E}{2}. \tag{14.35}$$

That is, as $T \to 0$, E and F approach the same limit. Note that this value of E is essentially the negative of the energy of sublimation of the crystal at 0 K and gives us

$$\frac{Nu_0}{2} \approx -(\Delta E_s)_0 \tag{14.36}$$

as the connection between the potential well depth and an experimentally observable property of the crystal.

Because E and F approach the same value as $T \to 0$, we have

$$S = \frac{E-F}{T} \to 0. \tag{14.37}$$

That is, all atoms are in their ground vibrational states, bound to their lattice points, and the macroscopic system has available to it only one microstate. Finally

$$C_V \to 3Nk\left(\frac{\Theta_E}{T}\right)^2 (e^{-\Theta_E/2T})^2, \tag{14.38}$$

implying that $C_V \to 0$ exponentially as $T \to 0$. As mentioned above, this rate of approach to zero is much more rapid than is observed experimentally.

At high temperatures, T large compared to Θ_E, we have

$$F \to \frac{Nu_0}{2} + 3NkT \ln\left(\frac{\Theta_E}{T}\right) \tag{14.39}$$

and

$$E \to \frac{Nu_0}{2} + \frac{3Nk\Theta_E}{2}\left(\frac{2T}{\Theta_E}\right)$$

$$\to \frac{Nu_0}{2} + 3NkT. \tag{14.40}$$

That is, each vibrational mode has associated with it the classical value of the energy, kT.

The entropy is

$$S \to 3Nk\left[\left(\frac{\Theta_E}{2T}\right)\left(\frac{2T}{\Theta_E}\right) - \ln\left(\frac{2\Theta_E}{2T}\right)\right]$$

$$\to 3Nk\left[1 + \ln\left(\frac{T}{\Theta_E}\right)\right], \tag{14.41}$$

and finally, we have for C_V,

$$C_V \to 3Nk\left(\frac{\Theta_E}{2T}\right)^2\left(\frac{2T}{\Theta_E}\right)^2$$

$$\to 3Nk, \tag{14.42}$$

the classical value corresponding to the empirically observed law of Dulong and Petit.

BIBLIOGRAPHY

D. L. Goodstein, *States of Matter*, Prentice-Hall, Englewood Cliffs, NJ, 1975, Chapter 3.

T. L. Hill, *Introduction to Statistical Thermodynamics*, Addison-Wesley, 1960, Chapter 5, Appendix D.

E. L. Knuth, *Introduction to Statistical Thermodynamics*, McGraw-Hill, New York, 1966, Chapter 9.

W. G. V. Rosser, *An Introduction to Statistical Physics*, Ellis Horwood, Chichester, UK, 1982, Chapter 10.

J. W. Whalen, *Molecular Thermodynamics*, John Wiley & Sons, New York, 1991, Chapter 7, Appendix D.

PROBLEMS

14.1 Calculate the relative contributions to u_0 from nearest, next-nearest, and third-nearest neighbors for the case of an FCC crystal for which the pair potential is the Lennard-Jones 6-12 potential.

14.2 Develop an expression for the vibrational frequency of an Einstein crystal as a function of ε, r_e, and c for the case where the pair potential is the Lennard-Jones 6-12 potential.

14.3 Develop expressions for P and for μ for the Einstein solid, considering u_0 and Θ_E to be functions of (V/N). Show that $(\partial F/\partial N)_{V,T}$ and $(F + PV)/N$ give the same expression for μ.

14.4 Calculate the change in Helmholz free energy accompanying the cooling of a cube of Al 1 cm on a side from 1000 K to 100 K. Assume $\Theta_E = 300$ K for Al.

14.5 Calculate the entropy of one mole of iron at 300 K and 0.1 MPa using the Einstein model. Assume $\Theta_E = 340$ K for Fe.

14.6 Develop an expression for the isothermal compressibility of an Einstein crystal.

CHAPTER 15

THE DEBYE MODEL OF THE SOLID

As already noted, the Einstein model of the crystalline solid leads to an expression for the specific heat which predicts a much too rapid approach to zero at low temperatures. This result is inherent in the model, which assumes that all of the vibrational modes of the crystal have the same frequency, and a fairly high frequency at that. In real crystals, the experimental evidence is that vibrational modes having a wide range of frequencies, and extending down to quite low frequencies, are available. In order to more realistically represent the behavior of real systems, then, we must develop a model that explicitly accounts for this wide range of vibrational mode frequencies.

THE DEBYE MODEL

One model that has been developed to do this is the so-called Debye model of the solid. This model represents the crystal as an elastic continuum in which the vibrational motions of the atoms give rise to three-dimensional standing waves having a range of frequencies. In order to be a stable standing wave in a crystal of mean linear dimension a, the wavelength of the wave is governed by the condition

$$\lambda = \frac{2a}{n} \qquad (15.1)$$

That is, there must be an integral number, n, of half-wave periods within the crystal. The longest available wavelength, λ_{max}, will be $2a$. Since the frequency

176 THE DEBYE MODEL OF THE SOLID

and wavelength are related by

$$v = \frac{c}{\lambda}, \tag{15.2}$$

where c is the velocity of propagation of the wave, this longest wavelength represents the minimum frequency and consequently the lowest-energy vibrational mode.

In order to develop an expression for the partition function for this system in terms of the energies associated with these allowed vibrational frequencies, we must have some means of evaluating the total range of allowed wavelengths. We will carry out this evaluation using an approach similar to that used in determining the number of translational modes having energies less than some particular value, ε. In the present case, for a one-dimensional model, the total number of wavelengths having a value longer than or equal to some particular value, λ, is

$$i = \frac{2a}{\lambda}. \tag{15.3}$$

If we generalize to the case of a three-dimensional standing wave, we can write similar expressions in terms of wavelength components along the principal axes of the crystal as

$$i_x = \frac{2a}{\lambda/\cos\theta_1}, \quad i_y = \frac{2a}{\lambda/\cos\theta_2}, \quad i_z = \frac{2a}{\lambda/\cos\theta_3}, \tag{15.4}$$

in which the $\cos\theta_i$ terms represent the angles between the crystal axes and the principal direction of the standing wave.

From this point, we may proceed in a manner analogous to that used in enumerating the allowed translational quantum states in the case of the ideal gas. Consider a three-dimensional quantum number space, with the axes defined—in this case, as i_x, i_y, i_z, the x, y, and z components of the allowed wavelengths. We can write an expression for the radius of a sphere in this quantum number space, as before, as

$$R^2 = (i_x^2 + i_y^2 + i_z^2)$$
$$= \left(\frac{4a^2}{\lambda^2}\right)(\cos^2\theta_1 + \cos^2\theta_2 + \cos^2\theta_3). \tag{15.5}$$

Because of the geometry of the system, the $\cos^2\theta_i$ term is unity. As in the translational case, the total number of wavelengths greater than or equal to some particular value of λ, $G(\lambda)$, is given by the volume of one octant of a sphere in this quantum number space, which in this case is

$$G(\lambda) = \frac{4\pi R^3}{3}\left(\frac{1}{8}\right)$$

$$= \left(\frac{4\pi}{3}\right)\left(\frac{4a^2}{\lambda^2}\right)^{3/2}\left(\frac{1}{8}\right)$$

$$= \left(\frac{4\pi}{3}\right)\left(\frac{a^3}{\lambda^3}\right). \tag{15.6}$$

This is the appropriate expression for the number of allowed wavelengths between λ_{max} and λ.

At this point, it must be recognized that each wavelength has associated with it three vibrational modes, one longitudinal and two transverse, all of which may have different velocities of propagation, c_i. We may thus rewrite the expression for the total number of allowed wavelengths in terms of the total number of allowed vibrational modes, assuming for now that $c_t \neq c_l$, but that $c_{t_1} = c_{t_2}$, and using $v = c/\lambda$ to determine that

$$G(v) = \left(\frac{4\pi}{3}\right)a^3 v^3 \left(\frac{2}{c_t^3} + \frac{1}{c_l^3}\right), \tag{15.7}$$

as the total number of vibrational modes having frequencies less than or equal to some particular value, v.

What now determines v_{max}, the maximum allowed value of v? There is a fundamental limitation on the total number of allowed vibrational modes, set by the criterion that the total may not exceed the number of degrees of freedom of the system. As we have seen before, this total is three per atom, or $3N$ for the system as a whole. Therefore

$$G(v_{max}) = 3N, \tag{15.8}$$

or

$$\left(\frac{4\pi}{3}\right)a^3 (v_{max})^3 \left(\frac{2}{c_t^3} + \frac{1}{c_l^3}\right) = 3N, \tag{15.9}$$

or

$$\frac{3N}{v_{max}^3} = \left(\frac{4\pi}{3}\right)a^3 \left(\frac{2}{c_t^3} + \frac{1}{c_l^3}\right). \tag{15.10}$$

Note that v_{max} for a given material will depend on c_t and c_l, which in turn depend on the binding energy of the crystal and the mass of the atoms that make up the crystal. These are the same parameters that defined the single vibrational frequency in the Einstein treatment.

Because we will need it shortly in the evaluation of the partition function of this system, we will develop an expression for the number of modes in a given frequency range—for example, from v to $v + dv$. This is essentially a density of states expression for the vibrational modes, similar to the expression for the density of states developed earlier for translational modes. The required density is simply

$$g(v) = \frac{d}{dv}[G(v)]$$

$$= \frac{d}{dv}\left[\left(\frac{4\pi}{3}\right)a^3\left(\frac{2}{c_t^3} + \frac{1}{c_l^3}\right)v^3\right], \qquad (15.11)$$

or, if we make the substitution

$$\frac{3N}{v_{max}^3} = \left(\frac{4\pi}{3}\right)a^3\left(\frac{2}{c_t^3} + \frac{1}{c_l^3}\right), \qquad (15.12)$$

we have

$$g(v) = \left(\frac{3N}{v_{max}^3}\right)\frac{d}{dv}(v^3)$$

$$= \left(\frac{9N}{v_{max}^3}\right)v^2 \qquad (15.13)$$

as the number of vibrational modes in a given range dv about some particular value of v.

It is of interest at this point to compare the results of the Debye model with those of the Einstein model, and with more exact treatments, in terms of the distribution of allowed frequencies. Figure 15.1 shows this comparison, plotted as $g(v)$ versus v. The Einstein case is represented by the single frequency, v_E, and the Debye treatment is represented by the parabolic relation given in equation (15.13).

EVALUATION OF THE PARTITION FUNCTION

We are now in a position to evaluate the partition function for the crystal based on the Debye model. Since we are again considering a system having a fixed volume, we will again work with the canonical ensemble partition function, Q. For this case, as with the Einstein model, we may represent Q as the product of two contributions,

$$Q = q_0^N \prod_v q_v. \qquad (15.14)$$

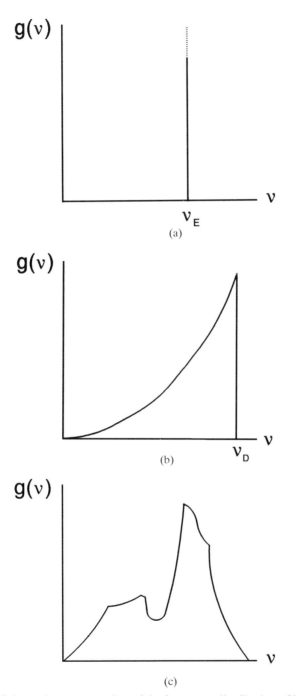

Figure 15.1 Schematic representation of the frequency distribution of lattice vibrations according to (a) the Einstein (single frequency) model, (b) the Debye model with $g(v) \propto v^2$, and (c) a more sophisticated model allowing for the effects of crystal structure.

180   THE DEBYE MODEL OF THE SOLID

Again as in the previous case, q_0 represents the contribution to the thermodynamic properties arising from the fact that the atoms of the crystal are bound to the other atoms, rather than being free. The value of q_0, again as in the previous case, is

$$q_0 = e^{-u_0/2kT}. \tag{15.15}$$

In this case, we can define Q_v, the contribution to Q from the $3N$ vibrational modes, as

$$Q_v = \prod_v q_v. \tag{15.16}$$

In this case, however, since all of the allowed modes have *different* frequencies, we cannot simply take one value for q_v and raise it to the $3N$ power. We must take the product of the $3N$ *different* values of q_v for each of the $3N$ allowed vibrational modes. (That is, the $3N$ vibrational modes are a set of *independent, different, distinguishable* subsystems, where the fact that they all have different frequencies makes them, in principle, distinguishable.) Rather than trying to take this product directly, we will deal with $\ln Q_v$, because this is the term that will appear in the expressions for the thermodynamic functions. Thus we have

$$\ln Q_v = \sum_v \ln q_v. \tag{15.17}$$

As we showed previously, we can write q_v for each allowed mode as

$$q_v = \frac{e^{-hv/2kT}}{1 - e^{-hv/kT}}, \tag{15.18}$$

or

$$\ln q_v = -\frac{hv}{2kT} - \ln(1 - e^{-hv/kT}). \tag{15.19}$$

Thus

$$\ln Q_v = \sum_v \left(-\frac{hv}{2kT} - \ln(1 - e^{-hv/kT}) \right). \tag{15.20}$$

Again, as in the calculation of q_t for the ideal gas, it would be more convenient if we could approximate this sum by an integral. As in the previous case, this is permissible if the energy differences between vibrational modes are small compared to kT. In the present case, the frequency increments between modes are of order

$$\Delta v \approx \frac{c}{2a}. \tag{15.21}$$

The ratio of the energy differences to kT is thus

$$\frac{\Delta \varepsilon}{kT} \approx \frac{h\Delta v}{kT} \approx \frac{hc}{2akT}. \tag{15.22}$$

If we assume that c is on the order of 10^5 cm/sec, a is on the order of 10 cm, and T is on the order of 300 K, we have $\Delta \varepsilon / kT$ on the order of 10^{-10}, which is small compared to unity. We may thus set up the equation for $\ln Q_v$ as an integral, analogously to the procedure used in the evaluation of q_t, as

$$\ln Q_v = \int g(v) \ln q_v \, dv$$

$$= \int_0^{v_{max}} \left(\frac{9N}{v_{max}^3} \right) v^2 \left[-\frac{hv}{2kT} - \ln(1 - e^{-hv/kT}) \right] dv, \tag{15.23}$$

to give, finally,

$$\ln Q = N \ln q_0 + \ln Q_v$$

$$= -\frac{Nu_0}{2kT} + \frac{9N}{v_{max}^3} \int_0^{v_{max}} \left[-\frac{hv}{2kT} - \ln(1 - e^{-hv/kT}) \right] v^2 \, dv$$

$$= -\frac{Nu_0}{2kT} - \frac{9Nh}{2kTv_{max}^3} \int_0^{v_{max}} v^3 \, dv - \frac{9N}{v_{max}^3} \int_0^{v_{max}} \frac{v^2}{\ln(1 - e^{-hv/kT})} \, dv$$

$$= -\frac{Nu_0}{2kT} - \frac{9Nhv_{max}}{8kT} - \frac{9N}{v_{max}^3} \int_0^{v_{max}} \frac{v^2}{\ln(1 - e^{-hv/kT})} \, dv. \tag{15.24}$$

EVALUATION OF THE THERMODYNAMIC FUNCTIONS

We will not evaluate the above integral at this point, but instead go directly to the expressions for the various thermodynamic functions. Proceeding as usual, we have

$$F = -kT \ln Q$$

$$= \frac{Nu_0}{2} + \frac{9Nhv_{max}}{8} + \frac{9NkT}{v_{max}^3} \int_0^{v_{max}} \frac{v^2}{\ln(1 - e^{-hv/kT})} \, dv. \tag{15.25}$$

182 THE DEBYE MODEL OF THE SOLID

$$E = kT^2 \left(\frac{\partial \ln Q}{\partial T}\right)_{N,V}$$

$$= kT^2 \left(\frac{Nu_0}{2kT^2}\right) + kT^2 \left(\frac{9Nhv_{max}}{kT^2}\right) - \frac{9NkT^2}{v_{max}^3} \frac{\partial}{\partial T} \left[\int_0^{v_{max}} \frac{v^2}{\ln(1 - e^{-hv/kT})} dv\right]$$

$$= \frac{Nu_0}{2} + \frac{9Nhv_{max}}{8} + \frac{9Nh}{v_{max}^3} \int_0^{v_{max}} \frac{v^3}{e^{hv/kT} - 1} dv. \tag{15.26}$$

$$S = \frac{E - F}{T}$$

$$= \frac{9Nh}{v_{max}^3 T} \int_0^{v_{max}} \frac{v^3}{e^{hv/kT} - 1} dv - \frac{9Nk}{v_{max}^3} \int_0^{v_{max}} \frac{v^2}{\ln(1 - e^{-hv/kT})} dv. \tag{15.27}$$

$$C_V = \left(\frac{\partial E}{\partial T}\right)_{N,V}$$

$$= \frac{9Nh}{v_{max}^3} \int_0^{v_{max}} v^3 \frac{\partial}{\partial T} (e^{hv/kT} - 1)^{-1} dv$$

$$= \frac{9Nh^2}{v_{max}^3 kT^2} \int_0^{v_{max}} \frac{v^4 e^{hv/kT}}{(e^{hv/kT} - 1)^2} dv. \tag{15.28}$$

At this point it is convenient to introduce the dummy variables

$$x = \frac{hv}{kT}, \qquad u = \frac{hv_{max}}{kT}. \tag{15.29}$$

Using these variables to substitute for v and v_{max} yields

$$F = \frac{Nu_0}{2} + \frac{9Nhv_{max}}{8} + \frac{9NkT}{u^3} \int_0^u \frac{x^2}{\ln(1 - e^{-x})} dx, \tag{15.30}$$

$$E = \frac{Nu_0}{2} + \frac{9Nhv_{max}}{8} + \frac{9NkT}{u^3} \int_0^u \frac{x^3}{e^x - 1} dx, \tag{15.31}$$

$$S = \frac{9Nk}{u^3} \left[\int_0^u \frac{x^3}{(e^x - 1)} dx + \int_0^u \frac{x^2}{\ln(1 - e^{-x})} dx\right], \tag{15.32}$$

$$C_V = \frac{9Nk}{u^3} \int_0^u \frac{x^4 e^x}{(e^x - 1)^2} dx. \tag{15.33}$$

The integrals in the above expressions cannot be solved in closed form. Consequently, it proves convenient to express all of the above functions in terms of a function known as the *Debye function*, which is defined as

$$D(u) = \frac{3}{u^3} \int_0^u \left(\frac{x^3}{e^x - 1}\right) dx = D\left(\frac{\Theta_D}{T}\right), \qquad (15.34)$$

where

$$\Theta_D \equiv \frac{h\nu_{max}}{k} \qquad (15.35)$$

is called the *Debye temperature*, by analogy with the previously defined Θ_E for the Einstein model. Values of $D(u)$ as a function of u have been calculated numerically and are tabulated in Appendix H, along with measured values of Θ_D for a number of materials. The final expression for the thermodynamic functions are thus

$$F = \frac{Nu_0}{2} + \frac{9Nk\Theta_D}{8} + 3NkT \ln(1 - e^{-\Theta_D/T}) - NkTD\left(\frac{\Theta_D}{T}\right), \qquad (15.36)$$

$$E = \frac{Nu_0}{2} + \frac{9Nk\Theta_D}{8} + 3NkTD\left(\frac{\Theta_D}{T}\right), \qquad (15.37)$$

$$S = 3Nk\left[\frac{4}{3}D\left(\frac{\Theta_D}{T}\right) - \ln(1 - e^{-\Theta_D/T})\right], \qquad (15.38)$$

$$C_V = 3Nk\left[4D\left(\frac{\Theta_D}{T}\right) - \frac{3\left(\frac{\Theta_D}{T}\right)}{(e^{\Theta_D/T} - 1)}\right], \qquad (15.39)$$

and finally

$$\mu = \frac{G}{N} \approx \frac{F}{N} = \frac{u_0}{2} + \frac{9k\Theta_D}{8} + 3kT \ln(1 - e^{-\Theta_D/T}) - kTD\left(\frac{\Theta_D}{T}\right). \qquad (15.40)$$

RELATION OF Θ_D TO CRYSTAL PROPERTIES

Let us look for a moment at the physical parameters on which Θ_D depends, in order to develop a feeling for the relative values of Θ_D to be expected for various substances. We may do this by substituting $k\Theta_D/h$ for ν_{max} in equation (15.12) to yield

$$\Theta_D = \left[\frac{9h^3}{4\pi k^3} \frac{N}{V} \frac{1}{\left(\frac{2}{c_t^3} + \frac{1}{c_l^3}\right)}\right]^{1/3}, \qquad (15.41)$$

where we have made the substitution $a^3 = V$. Note that Θ_D increases with increasing number density (atoms/unit volume) and with increasing velocities of propagation. These velocities in turn depend on the elastic modulus of the crystal, with high moduli leading to high velocities of propagation. Thus one would expect large values of Θ_D for materials with a high number density and high elastic modulus (i.e., a high resistance to elastic deformation), and vice versa. The classic example of a high-number-density, high-modulus material is diamond, for which $\Theta_D = 2230$ K. At the other end of the scale is lead, for which $\Theta_D = 88$ K. Θ_D values for a number of materials are summarized in Appendix H.

HIGH AND LOW TEMPERATURE LIMITS

As was the case with the Einstein treatment, it is instructive to look at the high and low temperature limits of these functions. To do so, we will begin by looking at the behavior of $D(u)$ at the limits of high and low temperatures. At low temperatures, as $T \to 0$, or as $u \to \infty$,

$$D(u) = \frac{3}{u^3} \int_0^u \left(\frac{x^3}{e^x - 1} \right) dx \to \left(\frac{\pi^4}{15} \right) \left(\frac{3}{u^3} \right). \tag{15.42}$$

At high temperature, as $T \to \infty$, or as u and $x \to 0$, the term in e^x may be expanded in a series to give

$$D(u) = \frac{3}{u^3} \int_0^u \left(\frac{x^3}{1 + x + \cdots - 1} \right) dx = \frac{3}{u^3} \int_0^u x^2 \, dx = 1. \tag{15.43}$$

Thus the expressions for E take the form

$$E \to \frac{Nu_0}{2} + \frac{9Nk\Theta_D}{8} + \frac{3Nk\Theta_D}{5} \left(\frac{T}{\Theta_D} \right)^4 \tag{15.44}$$

as $T \to 0$ and

$$E \to \frac{Nu_0}{2} + \frac{9Nk\Theta_D}{8} + 3NkT$$

$$\approx \frac{Nu_0}{2} + 3NkT, \tag{15.45}$$

(the second term on the right is small compared to the other two) as $T \to \infty$. We thus see that E approaches a fixed value of E_0 at low temperature, the rate of approach being proportional to T^4, and that E approaches the classical value of $E_0 + 3NkT$ at high temperatures, in agreement with the Einstein treatment and classical observation.

Finally, we may look at the high and low temperature limits of C_V, again using the limits on $D(u)$ deduced above. We determine that

$$C_V \to \frac{12Nk\pi^4}{5}\left(\frac{T}{\Theta_D}\right)^3 \qquad (15.46)$$

as $T \to 0$, and that

$$C_V \to 3Nk\left(4 - \frac{3u}{1 + u + \cdots - 1}\right) \to 4Nk, \qquad (15.47)$$

the classical value, as $T \to \infty$.

Again, as we did in the case of the Einstein Treatment, we may write

$$\frac{C_V}{Nk} = 3\left(4D\left(\frac{\Theta_D}{T}\right) - \frac{\frac{3\Theta_D}{T}}{e^{\Theta_D/T} - 1}\right). \qquad (15.48)$$

Thus we should be able to plot heat capacity data for all crystals on a reduced temperature plot of C_V/Nk versus T/Θ_D, as we did in the previous case. Such a plot, for a large number of substances, serves as a test of the accuracy of the Debye treatment, both on an absolute basis and in comparison with the Einstein treatment. Such a plot is shown in Figure 15.2. The Debye treatment,

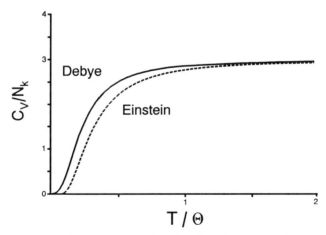

Figure 15.2 A reduced temperature plot showing the temperature variation of the constant volume heat capacity of crystalline solids according to the Debye model (solid line). The corresponding curve for the Einstein model is shown for comparison (dotted line).

which uses a much more realistic assumption concerning the available vibrational frequencies, gives a significantly better approximation to experimental observation.

BIBLIOGRAPHY

D. L. Goodstein, *States of Matter*, Prentice-Hall, Englewood Cliffs, NJ, 1975, Chapter 3.
T. L. Hill, *Introduction to Statistical Thermodynamics*, Addison-Wesley, 1960, Chapter 5.
C. C. Kittel and H. Kroemer, *Thermal Physics*, 2nd ed., W. H. Freeman & Co., San Francisco, 1980, Chapter 4.
E. L. Knuth, *Introduction to Statistical Thermodynamics*, McGraw-Hill, New York, 1966, Chapter 14.
W. G. V. Rosser, *An Introduction to Statistical Physics*, Ellis Horwood, Chichester, UK, 1982, Chapter 10.

PROBLEMS

15.1 Derive the low-temperature specific heat equation for a one-dimensional crystal, such as a fiber. Assume that vibrational displacements may occur both normal to the fiber axis and in the direction of the fiber chain. Take the total number of degrees of freedom to be $3N$. The definite integral

$$\int_0^\infty \frac{x}{e^x - 1} dx = \frac{\pi^2}{6}.$$

15.2 Would you expect the heat capacity at very low temperature to be the same for a large crystal ($\sim 1 \text{ cm}^3$) as for a very small crystallite ($\sim 3 \times 10^{-17} \text{ cm}^3$)? Explain your answer in terms of the Debye model of the solid. If you conclude that there would be a difference, which would have the larger value of C_V per mole at low temperature?

15.3 Calculate the change in internal energy when a 1-cm³ cube of copper is heated from 100 K to 800 K.
(a) Using the Debye treatment with $\Theta_D = 332$ K.
(b) Using the Einstein treatment, with $\Theta_E = 0.75 \Theta_D$.
Tables of the Debye function can be found in Appendix H.

15.4 Use the value of Θ_D given in Appendix H to calculate C_V per mole for Pb, Mg, and C (diamond) at 300 K.

15.5 Calculate Θ_D for aluminum, given that $C_V = 1.48 \text{ J/mol-K}$ at $T = 35.2$ K.

CHAPTER 16

SIMPLE LIQUIDS

We turn now to the problem of developing statistical mechanical expressions for the properties of simple liquids. We shall see that this is a difficult problem, because we must take into account interatomic interactions similar to those treated in the case of crystalline solids, but with the added complication that the liquid has, on a long-time basis, a random structure and relatively high atomic mobility.

THE MODEL

Many different models have been used to describe the behavior of liquids on a statistical mechanical basis. These models generally fall into one of two categories: (1) extrapolations of the treatment of nonideal gases, in which N/V is large enough that atomic size and interatomic interactions must be taken into account, giving rise to terms in the partition function related to these parameters, and (2) modifications to the treatments of crystalline solids to allow for the somewhat smaller N/V and increased atomic mobility. We will consider one of the latter models, known as a *cell model*, because it is relatively simple mathematically and provides a description of liquids at temperatures far removed from the critical temperature, where the density is only slightly lower than that of the corresponding solid.

The model that we will use considers an atom in the liquid to be bound in a potential well slightly larger than that in the corresponding Einstein model of the solid. This potential well is shown in one dimension in Figure 16.1, which may be compared to the corresponding potential well for the Einstein solid shown in Figure 14.1.

188 SIMPLE LIQUIDS

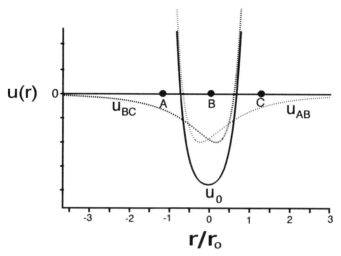

Figure 16.1 Potential well created about atom B due to summation of pairwise interactions with atom A and C in a linear chain, for the case of a "relaxed" Lennard-Jones 6-12 potential. The internuclear distance has been increased relative to the equilibrium distance arising from the Lennard-Jones formulation. The resulting value of u_0 is less than that shown in Figure 14.1 for the Einstein solid. The slope of the u_0 versus distance curve near $r/r_0 = 0$ is also reduced.

The potential energy of the liquid relative to the previously chosen zero of energy may be evaluated by the summation of pair potentials technique described in Chapter 14, with the additional complication that the r_{ij} are no longer fixed by having a periodic lattice of sites. We may again write

$$U(N) = \frac{1}{2} \sum_{i \neq j} u(r_{ij}), \tag{16.1}$$

for the potential energy contribution to system energy, and

$$u_0 = \sum_r u(r) n(r) \tag{16.2}$$

as the potential well depth. However, in this case we will not have integral numbers of neighbors at fixed distances, but some average number of atoms over the entire range of distances described by r. This average may be described in terms of the radial distribution function, $g(r)$, which describes the long-time average number of neighboring atoms per unit volume as a function of distance from a reference atom. A typical radial distribution function is shown in Figure 16.2. There is an initial maximum in the function at a value of r which is, in general, slightly larger than the nearest-neighbor distance in the corresponding crystalline solid phase, and lesser maxima at larger distances, finally dying out

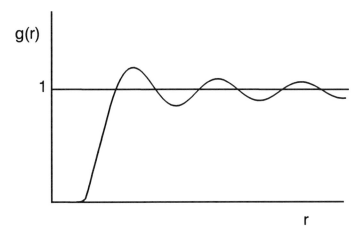

Figure 16.2 Schematic representation of the radial distribution function for a typical liquid, showing the long-time average number of neighbor atoms per unit volume at any distance, r, from the center of a reference atom.

to a constant value of unity at large distances. The expression for the potential well depth may be rewritten in terms of $g(r)$ as

$$u_0 = \frac{N}{V} \int u(r)g(r) \, d^3r. \tag{16.3}$$

Again as in the case of the crystalline solid, we may represent the total potential energy as $Nu_0/2$, where now $N_{\text{Av}}u_0/2$ is equal to $-(\Delta E_v)_0$, the energy that would be required to vaporize a mole of liquid at 0 K. (Keep in mind that although the free energy of the liquid at 0 K will be greater than that of the corresponding solid, the liquid could, in principle, exist as a metastable phase with well-defined thermodynamic properties at that temperature.) This potential energy will give rise to a term in the partition function similar to the corresponding term in the treatments of crystalline solids.

As usual, we must also take account of the energy involved in the motion of the atoms within the liquid. If the atoms of the liquid were able to move freely through the total volume of the liquid, as they do in the model of the ideal gas, we could account for the energy associated with this motion using a contribution to the partition function of the form

$$\frac{q_m^N}{N!} = \left(\frac{2\pi mkT}{h^2}\right)^{3N/2} \frac{V^N}{N!}. \tag{16.4}$$

In the case of the liqud, because of the presence of the other atoms, the atom may be considered to be confined to a "cell" of volume V/N, but able to move freely within this cell. If we adopt this description, divide the system up into an array of "cells," and assume for the moment that the cells are distinguishable, as were the sites in our models of crystalline solids, then we will have

$$q_m^N = \left(\frac{2\pi mkT}{h^2}\right)^{3N/2} \left(\frac{V}{N}\right)^N. \qquad (16.5)$$

If we now use Stirling's approximation on the $N!$ term in equation (16.4), we see that the two expressions above differ by a factor e^N. Because neither of the two models provides a physically exact description of the real state of affairs (the whole volume is certainly not available to each atom, and the atoms are not really distinguishable on a long-term basis, because they can move from cell to cell), there is no clear basis for choosing either of the above formulations. It has been common practice to retain the factor e^N, because it does, as we shall see later, lead to the correct order of magnitude for the entropy of fusion for atomic crystals. This contribution to q is often referred to as a *communal entropy*, arising from the loss of distinguishability that accompanies melting of the solid. This decision leads to

$$q_m^N = \left(\frac{2\pi mkT}{h^2}\right)^{3N/2} (v_{cell} e)^N. \qquad (16.6)$$

Finally, we must make one more correction to this formulation. The final problem that arises is that, because the atom itself occupies a volume comparable to the cell volume, the actual volume through which the atom can move is

$$v_f = v_{cell} - v_{atom}, \qquad (16.7)$$

where v_f is called the *free volume*.

EVALUATION OF THE PARTITION FUNCTION

We may now write out the complete expression for the partition function for this case as

$$\begin{aligned}Q &= q_0^N \cdot q_m^N \\ &= e^{-Nu_0/2kT} \left(\frac{2\pi mkT}{h^2}\right)^{3N/2} (v_f e)^N.\end{aligned} \qquad (16.8)$$

The only questions that now remain are those of how to evaluate u_0 and v_f. We will not address these questions at this point, but instead go directly to the expressions for the thermodynamic functions in terms of Q.

EVALUATION OF THE THERMODYNAMIC FUNCTIONS

Proceeding as usual, we obtain

$$F = -kT \ln Q$$
$$= \frac{Nu_0}{2} - NkT \ln\left[\left(\frac{2\pi mkT}{h^2}\right)^{3/2} (v_f e)\right]. \qquad (16.9)$$

The energy is

$$E = kT^2 \left(\frac{\partial \ln Q}{\partial T}\right)_{N,V}$$
$$= \frac{Nu_0}{2} + \frac{3}{2} NkT + NkT^2 \left(\frac{\partial \ln v_f}{\partial T}\right)_V. \qquad (16.10)$$

The heat capacity at constant volume is

$$C_V = \left(\frac{\partial E}{\partial T}\right)_{N,V}$$
$$= Nk\left[\frac{3}{2} + 2T\left(\frac{\partial \ln v_f}{\partial T}\right)_V + T^2\left(\frac{\partial^2 \ln v_f}{\partial T^2}\right)_V\right]. \qquad (16.11)$$

The entropy is

$$S = \frac{E - F}{T}$$
$$= Nk\left[\frac{5}{2} + \ln\left(\frac{2\pi mkT}{h^2}\right)^{3/2} v_f\right] + NkT\left(\frac{\partial \ln v_f}{\partial T}\right)_V. \qquad (16.12)$$

Finally, using the same logic as was used in the case of the Einstein crystal—namely, that the term PV is very small compared to F—we may write

$$\mu \approx \frac{F}{N} = \frac{u_0}{2} - kT \ln\left[\left(\frac{2\pi mkT}{h^2}\right)^{3/2} v_f\right] - kT. \qquad (16.13)$$

EVALUATION OF PARAMETERS

In order for the relations developed above to be of any use, we must evaluate the parameters u_0 and v_f. Evaluation of u_0 involves, in principle, the solution of equation (16.3), which, in turn, requires a knowledge of $g(r)$ and the appropriate pair potential. Alternatively, one can use the relation between u_0 and $-(\Delta E_v)_0$ mentioned earlier, provided that this parameter can be extrapolated from experimental data.

Evaluation of the term v_f is much more difficult, and it remains a central problem in the statistical thermodynamic treatment of liquids. We will discuss two relatively simple approaches to this problem, one of which will be of use in describing solid–liquid equilibrium at temperatures not too far removed from the triple point, and one of which will be of use in describing behavior closer to the critical point.

For the solid–liquid case, we may begin with a second definition of v_f, namely,

$$v_f = \int_{cell} e^{-(u_r - u_0)/kT}\, dr, \tag{16.14}$$

where in this case u_r is the value of u at distance r from r_0. This relation simply indicates that the volume available to the atom is constrained by the increase in system potential energy when the atom tries to move away from the center of the cell, and that the probability of finding the atom at distance r decreases exponentially with the increased potential energy associated with the required displacement from r_0. We further assume that the potential is that appropriate to a classical harmonic oscillator model, which is

$$u_r = \frac{2\pi v^2}{2m} r^2 dr. \tag{16.15}$$

This leads to

$$v_f = \left[\frac{1}{v}\left(\frac{kT}{2\pi m}\right)^{1/2}\right]^3. \tag{16.16}$$

Inserting this value into the expression for q_m yields

$$q_m = \left(\frac{2\pi m kT}{h^2}\right)^{3/2} \left[\frac{1}{v}\left(\frac{kT}{2\pi m}\right)^{1/2}\right]^3 e$$

$$= \left(\frac{kT}{hv}\right)^3 e. \tag{16.17}$$

In this expression, $(kT/hv)^3$ is simply the partition function for a three-dimensional harmonic oscillator at temperatures high enough that the oscillator behaves classically. It may be compared to the similar term that appears in the expression for Q for the Einstein solid for the case of $T \gg \Theta_E$. The resulting expressions for the thermodynamic functions are [assuming that

$(\partial v/\partial T)_V = 0$, as would be the case for a harmoic oscillator system at constant V/N]

$$F = \frac{Nu_0}{2} - 3NkT \ln\left(\frac{kT}{hv}\right) - NkT, \tag{16.18}$$

$$E = \frac{Nu_0}{2} + \frac{3}{2}NkT + \frac{3}{2}NkT$$

$$= \frac{Nu_0}{2} + 3NkT, \tag{16.19}$$

$$C_V = 3Nk, \tag{16.20}$$

$$S = 4Nk + 3Nk \ln\left(\frac{kT}{hv}\right), \tag{16.21}$$

and finally

$$\mu = \frac{u_0}{2} - kT \ln\left(\frac{kT}{hv}\right)^3 - kT. \tag{16.22}$$

Note that these expressions for F, E, C_V, and S have the same form as those developed in Chapter 14 for the Einstein crystal at high temperature, except for the additional factor of NkT in F and S, and kT in μ, arising from the "communal entropy" term.

An alternative approach to the evaluation of v_f is to take a geometrical approach and consider the problem in terms of a model in which the atoms are considered as hard spheres of diameter σ, which may be approximated by the collision diameter in the gas phase. If we consider each atom to be constrained to a spherical volume whose diameter is determined by the excluded volume associated with its nearest neighbors, assuming these neighbors to be fixed at their equilibrium positions, as shown in Figure 16.3, we arrive at

$$v_f = \left[2\left(\left(\frac{V}{N}\right)^{1/3} - \sigma\right)\right]^3. \tag{16.23}$$

For this case,

$$q_m = \left(\frac{2\pi mkT}{h^2}\right)^{3/2} [2(v^{1/3} - \sigma)]^3 e, \tag{16.24}$$

SIMPLE LIQUIDS

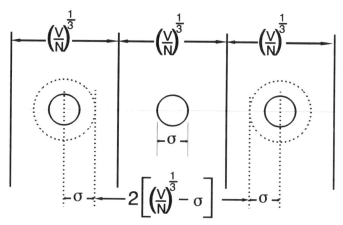

Figure 16.3 Schematic representation of the parameters used to define the free volume available to an atom in a liquid in terms of the atomic diameter, σ_e, and the volume per atom in the liquid, V/N.

where

$$v = \frac{V}{N}. \tag{16.25}$$

The corresponding expression for the chemical potential is

$$\mu = \frac{u_0}{2} - kT\left[\ln\left(\frac{2\pi mkT}{h^2}\right)^{3/2} + 3\ln[2(v^{1/3} - \sigma)]\right] - kT. \tag{16.26}$$

This model is better able to describe the liquid at temperatures closer to the critical point than is the expression based on the classical oscillator model, because the decrease in density at higher temperatures appears explicitly in the expression for chemical potential.

BIBLIOGRAPHY

D. L. Goodstein, *States of Matter*, Prentice-Hall, Englewood Cliffs, NJ, 1975, Chapter 11.

T. L. Hill, *Introduction to Statistical Thermodynamics*, Addison-Wesley, 1960, Chapter 20.

J. D. Hirschfelder, C. F. Curtis, and R. B. Bird, *Molecular Theory of Gases and Liquids*, John Wiley & Sons, New York, 1960, Chapter 4.

Y. Marcus, *Introduction to Liquid State Chemistry*, John Wiley & Sons, New York, 1977, Chapters 1–3.

J. W. Whalen, *Molecular Thermodynamics*, John Wiley & Sons, New York, 1991, Chapter 8.

PROBLEMS

16.1 Calculate the ratio of $(u_0)^L$ to $(u_0)^S$ and the ratio of v_l to v_s, assuming that the appropriate pair potential is the Lennard-Jones 6-12 potential, that the volume increase on melting is 10%, and that the form of the potential well for r close to r_0 is

$$u_r = u_0 + \left(\frac{\partial^2 u}{\partial r^2}\right)_0 (r - r_0)^2.$$

16.2 Calculate the chemical potential of silver at 1400 K, using the classical oscillator expression for q_m, and assuming that, to first order,

$$\frac{v_l}{v_E} = \left(\frac{\rho_l}{\rho_s}\right)^{1/3}.$$

16.3 Calculate the change in E associated with the heating of 1 cm^3 of Al from 1000 to 1200 K. You may use the expression for q_m based on the classical oscillator model, and assume that the oscillator frequencies for the solid and liquid are given by equations (17.21) and (17.24), respectively.

16.4 In Chapter 11, we stated that $\Delta S_f \approx 10$ J/mol-K for many metals. Explain this observation using the Einstein model for the solid and the classical oscillator model for the liquid at the melting point.

CHAPTER 17

STATISTICAL THERMODYNAMICS OF PHASE EQUILIBRIUM IN ONE-COMPONENT SYSTEMS

Now that we have developed models for one-component solids, liquids, and gases, we are in a position to use these models to describe phase equilibrium in single-component systems. That is, we have, in principle, the means of determining the equilibrium phase diagrams for these systems. In what follows, we will develop expressions for such phase diagram parameters as the conditions for solid–vapor solid–liquid, and vapor–liquid equilibrium, the triple point, and the boiling point.

SOLID–VAPOR EQUILIBRIUM

We will begin by developing an expression for the equilibrium vapor pressure over a solid phase of a pure material. This will prove to be the simplest case mathematically, and the one that the models we have developed are best able to treat realistically. We touched on this question in Chapter 11 when we developed the Clapeyron equation and applied it to solid–vapor and liquid–vapor equilibrium to obtain the relation

$$P^0 = Ke^{-\Delta H_i/RT}, \tag{17.1}$$

where at that point K was an undetermined constant and the ΔH_i term was the enthalpy of sublimation or vaporization. We may now use the statistical thermodynamic models that we have developed to calculate specific values for

P^0 as a function of temperature. For most systems of engineering interest, the pressure in the gas phase will be low enough that the gas phase can be adequately treated as an ideal gas. Whether the appropriate model is the monatomic or the polyatomic ideal gas will depend on the specific system being modeled. Typically, metallic systems have monatomic vapor phases at low pressures, although there are some exceptions such as As_4 and P_4.

The solid can be modeled by either the Einstein or the Debye model, or any more sophisticated model desired. For the present case, we will stick to the Einstein model, because the major features of the solid–vapor equilibrium are adequately modeled by this treatment.

The basis of the equilibrium calculation is the condition for distributive equilibrium developed in Chapter 5—namely, that the chemical potential must be uniform throughout the system at equilibrium. For the case of solid–vapor equilibrium, we have

$$\mu^S = \mu^V, \qquad (17.2)$$

or, putting in the appropriate expressions for μ^S for the Einstein solid [equation (14.33)] and for μ^V for the monatomic ideal gas [equation (12.45)],

$$\mu^S = \mu^V$$

$$\frac{u_0^S}{2} + 3kT \ln\left[2 \sinh\left(\frac{\Theta_E}{2T}\right)\right] = -kT \ln\left[\left(\frac{2\pi mkT}{h^2}\right)^{3/2} kT\right] + kT \ln P. \qquad (17.3)$$

This expression may be solved for the equilibrium vapor pressure as

$$\ln P^0 = \frac{u_0^S}{2kT} + 3 \ln\left[2 \sinh\left(\frac{\Theta_E}{2T}\right)\right] + \ln\left[\left(\frac{2\pi mk}{h^2}\right)^{3/2} k\right] + \frac{5}{2} \ln T. \qquad (17.4)$$

Once the atomic mass, the binding energy of the solid, and Θ_E are known, equation (17.4) may be used to determine the equilibrium vapor pressure at any temperature, and consequently the solid–vapor coexistence curve.

LIQUID–VAPOR EQUILIBRIUM

The procedure for characterizing liquid–vapor equilibrium is essentially the same as that presented above for solid–vapor equilibrium. The only question is that of which of the two expressions for μ^L developed in Chapter 16 is the appropriate one. Since most of the systems of engineering interest involve temperatures closer to the triple point than to the critical point, we will use the expression

$$\mu^L = \frac{u_0^L}{2} - kT \ln\left(\frac{kT}{hv_l}\right)^3 - kT, \qquad (17.5)$$

leading to the expression for the equilibrium vapor pressure over the liquid as

$$\ln P^0 = \frac{u_0^L}{2kT} - 3\ln\left(\frac{kT}{hv_l}\right)^3 - 1 + \ln\left[\left(\frac{2\pi mk}{h^2}\right)^{3/2} k\right] + \frac{5}{2}\ln T$$

$$= \frac{u_0^L}{2kT} + \ln\left[\left(\frac{2\pi mk}{h^2}\right)^{3/2} k\left(\frac{hv_l}{k}\right)^3 \frac{1}{e}\right] - \frac{1}{2}\ln T. \qquad (17.6)$$

Note that the form of this expression differs from that developed for the vapor pressure over the solid only in the presence of the additive term $kT \ln e$, associated with the concept of communal entropy introduced in Chapter 16 and in the constant multiplying the $\ln T$ term.

This last difference is more apparent than real. If we modify the expression developed for the equilibrium vapor pressure over the solid [equation (17.4)] to take account of the fact that T is generally greater than Θ_E at the melting point, and recall that $\Theta_E \equiv hv_E/k$, we arrive at

$$\ln P^0 = \frac{u_0^S}{2kT} + 3\ln\left(\frac{T}{\Theta_E}\right) + \ln\left[\left(\frac{2\pi mk}{h^2}\right)^{3/2} k\right] + \frac{5}{2}\ln T$$

$$= \frac{u_0^S}{2kT} + \ln\left[\left(\frac{2\pi mk}{h^2}\right)^{3/2} k\left(\frac{hv_E}{k}\right)^3\right] - \frac{1}{2}\ln T \qquad (17.7)$$

for the solid, which shows the same temperature dependence as the expression for the vapor pressure over the liquid. The values of u_0^L and v_l will, of course, differ from the corresponding values for the solid.

The above results may be compared to an empirical expression often used to relate P^0 to temperature, in order to show the correspondence between the terms occurring in the two equations. The empirical expression is of the form

$$\log_{10} P^0 = \frac{a}{T} + b\log_{10} T + cT + d. \qquad (17.8)$$

Comparison of equations (17.6) and (17.7) to equation (17.8) leads to the correspondences

$$a = \frac{u_0}{2k}, \qquad b = -(2.303)\frac{1}{2},$$

$$d = 2.303 \log_{10}\left[\left(\frac{2\pi mk}{h^2}\right)^{3/2} k\left(\frac{hv}{k}\right)^3\right]. \qquad (17.9)$$

(The c term in the empirical expression reflects the effect of factors not taken into account in the models used, and it is often zero.) These values may be compared to those shown in Appendix D, a compilation of results from empirical observation of a wide range of systems.

THE TRIPLE POINT

The triple point represents the point where all three phases are in mutual equilibrium and where the solid–vapor, liquid–vapor, and solid–liquid coexistence curves all intersect. Because we have already developed expressions for the solid–vapor and liquid–vapor coexistence curves, we may equate these expressions to determine the triple point using

$$(P^0)^S = (P^0)^L$$

$$\frac{u_0^S}{2kT_t} + \ln\left[\left(\frac{2\pi mk}{h^2}\right)^{3/2} k\left(\frac{hv_E}{k}\right)^3\right] - \frac{1}{2}\ln T_t = \frac{u_0^L}{2kT_t} + \ln\left[\left(\frac{2\pi mk}{h^2}\right)^{3/2} k\left(\frac{hv_l}{k}\right)^3\right]$$

$$\frac{u_0^S}{2kT_t} + 3\ln v_E = \frac{u_0^L}{2kT_t} + 3\ln v_l - \ln e, \qquad (17.10)$$

to yield

$$T_t = \frac{\dfrac{u_0^L - u_0^S}{2}}{k\left(3\ln\left(\dfrac{v_E}{v_l}\right) + 1\right)}. \qquad (17.11)$$

Once T_t has been established by this procedure, P_t, the triple point pressure, may be obtained by solving equation (17.8) or (17.9) for the equilibrium vapor pressure at T_t.

Note that we could have arrived at a general expression for the melting temperature, having the same form as equation (17.11), by directly equating the chemical potentials in the solid and liquid phases. The equation developed above implicitly implies that the values of the potentials and vibrational frequencies are those appropriate to the triple point pressure, while the general expression for the melting point would require the appropriate values of these parameters for whatever the applied pressure was in the case of interest.

SOLID–LIQUID EQUILIBRIUM

We have already touched on the subject of solid–liquid equilibrium in our discussion of the triple point above. The sole remaining question that must be resolved in the determination of the one-component phase diagram is that of the solid–liquid coexistence line. One approach to this question would be to use the expression developed above for the triple point and to develop explicitly the dependence of the u_0 and v terms on pressure. As a practical matter, this would involve making assumptions about (a) the compressibility and thermal expansion coefficients of the two phases and (b) the way in which these terms vary with pressure and temperature. This process would involve modeling small differences between large numbers, a process likely to result in large errors. Instead, we will make the assumption that the change of the melting point with pressure over the range of practical interest is essentially linear and can be obtained by applying the Clapeyron equation in the vicinity of the triple point. This leads to

$$\frac{dP}{dT} = \frac{\Delta S_f}{\Delta V_f} = \frac{S^L - S^S}{M\left(\frac{1}{\rho_l} - \frac{1}{\rho_s}\right)}, \quad (17.12)$$

where we may obtain the S^i from equations (14.41) and (16.21), M is the molecular weight of the species involved, and the ρ_i are the appropriate densities. Substitution of the expressions for the molar entropies yields

$$\frac{dP}{dT} = \frac{4R + 3R\ln\left(\frac{kT}{hv_l}\right) - 3R - 3R\ln\left(\frac{kT}{hv_s}\right)}{M\left(\frac{1}{\rho_l} - \frac{1}{\rho_s}\right)}$$

$$= \frac{R}{M} \frac{1 + 3\ln\left(\frac{v_s}{v_l}\right)}{\frac{1}{\rho_l} - \frac{1}{\rho_s}}. \quad (17.13)$$

Note in passing that, for the model chosen,

$$\Delta S_f = R\left(1 + 3\ln\left(\frac{v_s}{v_l}\right)\right). \quad (17.14)$$

We will return to this expression when we consider phase equilibria in multicomponent systems in Chapter 23.

A NUMERICAL EXAMPLE

We are now in position to construct the phase diagram for a typical one-component system, provided that we know the molecular weight of the species involved, that we know whether the gas phase is monatomic or polyatomic, and that we can model the potential energy and vibrational frequency terms.

202 STATISTICAL THERMODYNAMICS OF PHASE EQUILIBRIUM

We will begin this process by assuming that we can describe the system using a nearest-neighbor model, with pairwise additivity of interatomic potentials. We will assume further that we may use the Lennard-Jones 6-12 potential to describe the pairwise interatomic interaction. Note that this is an arbitrary decision—we could equally well use any other model potential. The Lennard-Jones potential has been chosen because it is relatively simple mathematically. Subject to these assumptions, we may express the binding energy term for the solid phase as

$$u_0^S = \sum_r u(r)n(r) \tag{17.15}$$

in general, or for the case of a crystalline solid with nearest-neighbor interactions only:

$$u_0^S = cu(r_0) = c\varepsilon\left(\frac{r_e^{12}}{r_0^{12}} - 2\frac{r_e^6}{r_0^6}\right) \tag{17.16}$$

where c is the coordination number in the solid.

We may relate r_0 to r_e by taking the derivative of u_0 with respect to r and setting this derivative equal to zero at $r = r_0$. This process yields

$$\left(\frac{du_0^S}{dr}\right)_{r=r_0} = c\varepsilon\left(\frac{-12r_e^{12}}{r_0^{13}} + \frac{12r_e^6}{r_0^7}\right) = 0, \tag{17.17}$$

leading to

$$r_0 = r_e, \quad u_0^S = -c\varepsilon. \tag{17.18}$$

This will be a general result for any solid that can be described by the Lennard-Jones potential and nearest-neighbor interactions only.

We may obtain the expression for v_s in terms of ε and r_0 by recalling that in the harmonic approximation

$$v = \frac{1}{2\pi}\left(\frac{f}{m}\right)^{1/2}, \tag{17.19}$$

where

$$f = \left(\frac{\partial^2 u_0}{\partial r^2}\right)_{r=r_0} = c\varepsilon\left(\frac{(12 \times 13)r_e^{12}}{r_0^{14}} - \frac{(12 \times 7)r_e^6}{r_0^8}\right)$$

$$= c\varepsilon\left(\frac{72}{r_0^2}\right), \tag{17.20}$$

as $r_0 = r_e$. From this we may obtain v_s as

$$v_s = \frac{1}{2\pi}\left[\frac{c\varepsilon}{m}\left(\frac{72}{r_{0_s}^2}\right)\right]^{1/2} = \frac{3\sqrt{2}}{\pi}\left(\frac{c\varepsilon}{mr_{0_s}^2}\right)^{1/2}. \qquad (17.21)$$

Thus a knowledge of the crystal structure of the solid, the values of r_0 and ε, or equivalently the density and energy of sublimation of the solid, are sufficient to determine both u_0^S and v_s.

The evaluation of u_0^L and v_l in terms of the model chosen requires additional assumptions. Ideally, we would require the radial distribution function for the liquid in order to determine u_0^L. Lacking this, we will assume that we can treat the liquid using the same pair potential as we used for the solid, with the same average coordination number, but with an interatomic distance consistent with the density of the liquid. Again, there are other assumptions that could be made to relate the properties of the liquid to those of the solid. The assumptions made here have been chosen, as before, to provide mathematical simplicity while leading to a result that retains the major features of the behavior of real systems.

Subject to these assumptions, we may write

$$u_0^L = c\varepsilon\left(\frac{r_e^{12}}{r_{0_l}^{12}} - 2\frac{r_e^6}{r_{0_l}^6}\right), \qquad (17.22)$$

where now $r_{0_l} \neq r_e$ and can be obtained from the density of the liquid. Similarly, the expression for f_l is

$$f_l = \left(\frac{\partial^2 u_0^L}{\partial r^2}\right)_{r=r_{0_l}} = c\varepsilon \frac{\partial^2}{\partial r^2}\left(\frac{r_e^{12}}{r^{12}} - 2\frac{r_e^6}{r^6}\right)_{r=r_{0_l}}$$

$$= \frac{c\varepsilon}{r_{0_l}^2}\left[156\left(\frac{r_e}{r_{0_l}}\right)^{12} - 84\left(\frac{r_e}{r_{0_l}}\right)^6\right], \qquad (17.23)$$

leading to

$$v_l = \frac{1}{2\pi}\left\{\frac{c\varepsilon}{mr_{0_l}^2}\left[156\left(\frac{r_e}{r_{0_l}}\right)^{12} - 84\left(\frac{r_e}{r_{0_l}}\right)^6\right]\right\}^{1/2}. \qquad (17.24)$$

We may now proceed to the numerical example. We will assume that the system is made up of atoms of molecular weight $M = 50$, that

$$\rho_s = 5.00 \text{ g/cm}^3, \qquad \rho_l = 4.20 \text{ g/cm}^3, \qquad \varepsilon = 40000 \text{ J/mol},$$

that the solid has the face-centered cubic crystal structure, for which $c = 12$, and that the gas phase is monatomic. The value of u_0^S is thus

$$u_0^S = -12\varepsilon = -480 \text{ kJ/mol}. \qquad (17.25)$$

204   STATISTICAL THERMODYNAMICS OF PHASE EQUILIBRIUM

For the face-centered cubic crystal structure, r_{0_s} is related to ρ_s by

$$\rho_s = \frac{4M}{N_{Av}} \left(\frac{\sqrt{2}}{2r_{0_s}}\right)^3, \tag{17.26}$$

leading to

$$r_{0_s} = 2.87 \times 10^{-8} \text{ cm}. \tag{17.27}$$

We may use this value of r_{0_s} to calculate v_s as

$$v_s = \frac{3\sqrt{2}}{\pi} \left(\frac{c\varepsilon}{mr_{0_s}}\right)^{1/2} = 1.5 \times 10^{13} \text{ sec}^{-1}, \tag{17.28}$$

and from this

$$\Theta_E = \frac{hv_s}{k} = 698 \text{ K}. \tag{17.29}$$

The value of u_0^L may be determined from equation (17.22), recalling that $r_e = r_{0_s}$ and that, from the way the model was defined,

$$\frac{r_{0_s}}{r_{0_l}} = \left(\frac{\rho_l}{\rho_s}\right)^{1/3}, \tag{17.30}$$

leading to

$$u_0^L = c\varepsilon \left[\left(\frac{\rho_l}{\rho_s}\right)^4 - 2\left(\frac{\rho_l}{\rho_s}\right)^2\right] = -438 \text{ kJ/mol}. \tag{17.31}$$

Finally, v_l is given by

$$v_l = \frac{1}{2\pi} \left\{\frac{c\varepsilon}{mr_{0_l}^2} \left[156\left(\frac{\rho_l}{\rho_s}\right)^4 - 84\left(\frac{\rho_l}{\rho_s}\right)^2\right]\right\}^{1/2} = 7.0 \times 10^{12} \text{ sec}^{-1} \tag{17.32}$$

The pertinent features of the phase diagram may now be evaluated. The triple point is given by equation (17.11) as

$$T_t = \frac{-438{,}000 - (-480{,}000)}{(2)(8.314)\left[3\ln\left(\frac{1.5 \times 10^{13}}{7 \times 10^{12}}\right) + 1\right]} = 768 \text{ K}. \tag{17.33}$$

The solid–vapor coexistence line, for temperatures below T_t, may be calculated from equation (17.4) as

$$\ln(P^0)^S = \frac{-480{,}000}{(2)(8.314)T} + 3\ln\left[2\sinh\left(\frac{698}{2T}\right)\right]$$

$$+ \ln\left[\left(\frac{2\pi(50)(1.38\times10^{-16})}{(6\times10^{23})(6.6\times10^{-27})^2}\right)^{3/2}\frac{82.05}{6\times10^{23}}\right] + \frac{5}{2}\ln T$$

$$= \frac{-28{,}867}{T} + \ln\left[2\sinh\left(\frac{349}{T}\right)\right]^3 + \ln 9.328 + \frac{5}{2}\ln T, \qquad (17.34)$$

or

$$(P^0)^S = 2.22\left[2\sinh\left(\frac{349}{T}\right)\right]^3 T^{5/2} e^{-28867/T} \text{ atm}. \qquad (17.35)$$

The liquid–vapor coexistence line, for temperatures above T_t, may be calculated from equation (17.6) as

$$\ln(P^0)^L = \frac{-438{,}000}{(2)(8.314)T} + \ln\left(\frac{(6.6\times10^{-27})(7\times10^{12})}{1.38\times10^{-16}}\right)^3 + \ln 9.238$$

$$= \frac{-26{,}341}{T} + \ln(337)^3 + 2.22 - \frac{1}{2}\ln T$$

$$= \frac{-26{,}341}{T} + 19.68 - \frac{1}{2}\ln T, \qquad (17.36)$$

or

$$(P^0)^L = 19.68\, T^{-1/2} e^{-26{,}341/T} \text{ atm}. \qquad (17.37)$$

Finally, we may construct the solid–liquid coexistence curve as a line extending from the triple point, having a slope given by equation (17.13) as

$$\left(\frac{dP}{dT_m}\right) = \frac{82.05}{50} \frac{\left(1 + 3\ln\dfrac{1.5\times10^{13}}{7\times10^{12}}\right)}{\dfrac{1}{4.2} - \dfrac{1}{5.0}} = 142 \text{ atm/K}. \qquad (17.38)$$

The resulting phase diagram is shown in Figure 17.1, plotted as $\ln P$ versus T. It can be seen from the figure that this diagram shows all of the features expected for equilibrium in a one-component system, and that the values obtained for parameters such as the melting and boiling points are physically

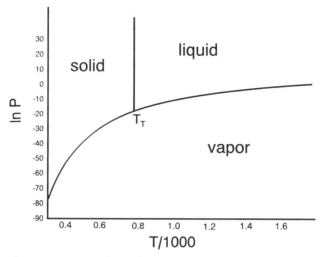

Figure 17.1 One-component phase diagram for a hypothetical system in which the properties of the solid, liquid, and vapor phases are described by the models developed in Chapters 14, 16, and 12, respectively.

realistic. It should be borne in mind, however, that the ability to model the behavior of any real system depends on (a) the availability of a realistic model for the interatomic potential and (b) a valid procedure for determining the binding energy and vibrational frequency terms using this potential. The value of the procedure outlined above is primarily in (a) identifying the parameters of importance in describing equilibrium behavior and (b) demonstrating the way in which changes in these parameters affect that equilibrium behavior.

BIBLIOGRAPHY

P. Gordon, *Principles of Phase Diagrams in Materials Systems*, McGraw-Hill, New York, 1968, Chapter 3.

C. C. Kittel and H. Kroemer, *Thermal Physics*, 2nd Ed., W. H. Freeman & Co., San Francisco, 1980, Chapter 10.

J. W. Whalen, *Molecular Thermodynamics*, John Wiley & Sons, New York, 1991, Chapters 9 and 10.

PROBLEMS

17.1 The equilibrium vapor pressure of solid aluminum at 815 K is 1×10^{-11} torr (mm of Hg). Calculate the molar enthalpy of sublimation, ΔH_s, assuming that the solid behaves as an Einstein solid with

$\Theta_E = 300\,\text{K}$ and that the vapor behaves as a monatomic ideal gas. Assume that $\omega_{e_1} = 1$.

$$1\,\text{torr} = 1333\,\text{dyn/cm}^2 = 13.3\,\text{kg/m}^2 = 13.3\,\text{Pa}$$

17.2 Calculate the vapor pressure of copper at 800 K:
  (a) By using thermodynamic data from Appendix D.
  (b) By using statistical thermodynamics, assuming that the vapor is a monatomic ideal gas and that the crystal follows the Debye model.

State any additional assumptions that you make.

17.3 Calculate the molar entropy of sublimation of aluminum at 100 K at a pressure of 0.1 MPa, assuming that the solid can be treated as an Einstein crystal with $\Theta_E = 300\,\text{K}$ and that the vapor can be treated as a monatomic ideal gas.

17.4 Calculate the melting point of silver using the equations developed in this chapter and compare the result to the measured value given in Appendix C.

17.5 Calculate the rate of change of the melting point of copper with pressure:
  (a) Using the Clapeyron equation and the thermodynamic data from Appendix C.
  (b) Using equations developed in this chapter.

PART IV

MULTICOMPONENT SYSTEMS

CHAPTER 18

CLASSICAL THERMODYNAMICS OF MULTICOMPONENT SYSTEMS

We turn now to the consideration of systems containing more than one component. As in the previous case of one-component systems, we will begin by developing the classical thermodynamic expressions that describe multicomponent systems, then develop the statistical mechanical framework required to evaluate these expressions for various model systems.

ACTIVITY

We will begin by making an additional definition, and then developing some concepts of general utility in the study of multicomponent systems. We will define the *activity* of a component i in a multicomponent system by the expression

$$d\mu_i = kT\, d\ln a_i, \tag{18.1}$$

which can be integrated to yield

$$\mu_i = \mu_i^0 + kT \ln a_i, \tag{18.2}$$

where, at this point μ_i^0 is an integration constant whose value is arbitrary. It will be convenient to define μ_i^0 in terms of the chemical potential of the pure component in whatever we chose to be its standard state at the temperature of

211

interest. That is,

$$\mu_i^0 = \mu \text{ of pure } i \text{ in its standard state} = kT \ln a_i^0. \tag{18.3}$$

The utility of this definition can be seen if we recall the expressions for μ_i developed in our study of the ideal monatomic gas. In that case we deduced that

$$\mu_i = -kT \ln\left[\left(\frac{2\pi m kT}{h^2}\right)^{3/2}\left(\frac{kT}{P_i}\right)\right]$$

$$= -kT \ln\left[\left(\frac{2\pi m kT}{h^2}\right)^{3/2} kT\right] + kT \ln P_i$$

$$= -kT \ln\left[\left(\frac{2\pi m kT}{h^2}\right)^{3/2}\left(\frac{kT}{P_i^0}\right)\right] + kT \ln\left(\frac{P_i}{P_i^0}\right)$$

$$= \mu_i^0 + kT \ln\left(\frac{P_i}{P_i^0}\right), \tag{18.4}$$

where

$$\mu_i^0 = -kT \ln\left[\left(\frac{2\pi m kT}{h^2}\right)^{3/2}\left(\frac{kT}{P_i^0}\right)\right]. \tag{18.5}$$

Comparison of this expression with equation (18.2) yields

$$a_i = \frac{P_i}{P_i^0} \tag{18.6}$$

for the case of an ideal gas. The reference pressure, P_i^0, can be arbitrarily chosen. Customarily it is chosen as the pressure at which the gas is in equilibrium with the pure condensed phase of the species of interest at the temperature of interest. That is, P_i^0 is the equilibrium vapor pressure over the condensed phase of the pure species i. Note that the final relation shown above applies *only* to the case of an ideal gas. In other cases the form may be quite different.

MOLAR PROPERTIES

Let us recall at this point, as an additional tool for describing the behavior of multicomponent systems, the concepts of mole fraction and molar properties developed in Chapter 11. We defined the mole fraction of component i in a system as

$$X_i = \frac{n_i}{\sum_i n_i}, \tag{18.7}$$

and correspondingly defined the molar values of the various extensive thermodynamic properties describing a given system as

$$S = \frac{S}{\sum_i n_i}, \quad G = \frac{G}{\sum_i n_i}, \quad V = \frac{V}{\sum_i n_i}, \tag{18.8}$$

and similarly for all of the other *extensive* thermodynamic properties. These relations apply to the system as a whole and give the values of S, G, and so on, for one mole of *total system* at constant compositon.

PARTIAL MOLAR QUANTITIES

We may also consider the contribution to the properties of the system made by each of the components that make up the system. Recall that we showed some time ago that

$$G = \sum_i N_i \mu_i = N_1 \mu_1 + N_2 \mu_2 + \cdots. \tag{18.9}$$

If we put this on a molar basis, we have

$$G = \sum_i \left(\frac{n_i}{\sum_i n_i}\right) N_{Av} \mu_i$$

$$= X_1 N_{Av} \mu_1 + X_2 N_{Av} \mu_2 + \cdots$$

$$= \sum_i \left(\frac{n_i}{\sum_i n_i}\right) \bar{G}_i$$

$$= X_1 \bar{G}_1 + X_2 \bar{G}_2 + \cdots, \tag{18.10}$$

where

$$\bar{G}_i = N_{Av} \mu_i. \tag{18.11}$$

Thus we see that the total molar Gibbs free energy of the system is simply the weighted sum of the $\bar{G}_i$, the partial molar Gibbs free energies of the components that make up the system, and that the partial molar Gibbs free energy is simply the chemical potential multiplied by Avogadro's number, N_{Av}. That is, we can think of each component as having a partial molar Gibbs free energy associated with it, given by

$$\bar{G}_i = N_{Av}\mu_i = \left(\frac{\partial G}{\partial n_i}\right)_{P,T,n_{j\neq i}}, \qquad (18.12)$$

which will be a function of T, P, and the other components present in the system. By similar reasoning we may develop expressions for other partial molar quantities. For example, we can write the total system volume as

$$V = n_1 \bar{V}_1 + n_2 \bar{V}_2 + \cdots \qquad (18.13)$$

or, in terms of one mole of system, as

$$V = X_1 \bar{V}_1 + X_2 \bar{V}_2 + \cdots, \qquad (18.14)$$

where

$$\bar{V}_i = \left(\frac{\partial V}{\partial n_i}\right)_{P,T,n_{j\neq i}}, \qquad (18.15)$$

and similarly for $\bar{S}_i$, $\bar{H}_i$, and $\bar{E}_i$.

If we now consider the changes that accompany a change in system composition at constant T and P, we can develop an additional useful relationship among the partial molar properties of the system. Recall that we developed the Gibbs–Duhem equation by starting from the expression

$$G = \sum_i N_i \mu_i, \qquad (18.16)$$

differentiating it to yield

$$dG = \sum_i N_i d\mu_i + \sum_i \mu_i dN_i \qquad (18.17)$$

and comparing this result to the relation

$$dG = -SdT + VdP + \sum_i \mu_i dN_i \qquad (18.18)$$

to yield

$$SdT - VdP + \sum_i N_i d\mu_i = 0. \tag{18.19}$$

If we limit ourselves to changes occurring at constant temperature and pressure, we have

$$\sum_i N_i d\mu_i = 0, \tag{18.20}$$

or, in terms of molar properties,

$$\sum_i X_i d\bar{G}_i = 0 \tag{18.21}$$

for any change in a multicomponent system about a point of internal equilibrium at constant P and T. We could equally well start with the expression for any of the other extensive properties in terms of partial molar properties and deduce similar relations. For example, for the case of volumes we have

$$V = \sum_i n_i \bar{V}_i, \tag{18.22}$$

or

$$dV = \sum_i n_i d\bar{V}_i + \sum_i \bar{V}_i dn_i. \tag{18.32}$$

This may be compared to the general expression

$$dV = \left(\frac{\partial V}{\partial T}\right)_{P, n_i} dT + \left(\frac{\partial V}{\partial P}\right)_{T, n_i} dP + \sum_i \left(\frac{\partial V}{\partial n_i}\right)_{P, T} dn_i, \tag{18.24}$$

to yield

$$\left(\frac{\partial V}{\partial T}\right)_{P, n_i} dT + \left(\frac{\partial V}{\partial P}\right)_{T, n_i} dP + \sum_i n_i d\bar{V}_i = 0, \tag{18.25}$$

the analog of the Gibbs–Duhem equation for partial molar volumes. As was the case for $\bar{G}_i$, if we have constant P and T, we may write

$$\sum_i n_i \bar{V}_i = 0, \tag{18.26}$$

or, in terms of mole fractions,

$$\sum_i X_i \bar{V}_i = 0, \tag{18.27}$$

and similarly for the other extensive thermodynamic properties.

RELATION OF PARTIAL TO TOTAL MOLAR PROPERTIES

We may go on to develop another set of useful relationships among the partial molar properties, based on their relationship to the total properties. We know from the defining relation for Gibbs free energy that

$$G = H - TS, \tag{18.28}$$

and from this

$$dG = dH - TdS - SdT. \tag{18.29}$$

If we take a system and change its composition by adding a small amount, dn_i, of some *one component*, i, at constant T, P, and $n_{j \neq i}$ we will have

$$dG = \left(\frac{\partial G}{\partial n_i}\right)_{T,P,n_{j \neq i}} dn_i, \quad dH = \left(\frac{\partial H}{\partial n_i}\right)_{T,P,n_{j \neq i}} dn_i, \quad dS = \left(\frac{\partial S}{\partial n_i}\right)_{T,P,n_{j \neq i}} dn_i, \tag{18.30}$$

or

$$\left(\frac{\partial G}{\partial n_i}\right)_{T,P,n_{j \neq i}} dn_i = \left(\frac{\partial H}{\partial n_i}\right)_{T,P,n_{j \neq i}} dn_i - T\left(\frac{\partial S}{\partial n_i}\right)_{T,P,n_{j \neq i}} dn_i, \tag{18.31}$$

or, dividing through by dn_i,

$$\left(\frac{\partial G}{\partial n_i}\right)_{T,P,n_{j \neq i}} = \left(\frac{\partial H}{\partial n_i}\right)_{T,P,n_{j \neq i}} - T\left(\frac{\partial S}{\partial n_i}\right)_{T,P,n_{j \neq i}}, \tag{18.32}$$

or

$$\bar{G}_i = \bar{H}_i - T\bar{S}_i. \tag{18.33}$$

That is, we have the same relation among the partial molar quantities associated with any one component that we have for the extensive properties of the system as a whole. Other similar relations follow from applying similar reasoning to the other appropriate relations among the total properties of the system.

We may deduce one final set of useful relations by looking at the partial derivatives obtained by differentiating these relations among the partial molar properties, such as

$$\left(\frac{\partial \bar{G}_i}{\partial T}\right)_{P, n_i} = -\bar{S}_i$$

$$\left[\frac{\partial(\bar{G}_i - G_i^0)}{\partial T}\right]_{P, n_i} = -(\bar{S}_i - S_i^0), \tag{18.34}$$

$$\left[\frac{\partial\left(\frac{\bar{G}_i}{T}\right)}{\partial \frac{1}{T}}\right]_{P, n_i} = \bar{H}_i$$

$$\left[\frac{\partial \frac{(\bar{G}_i - G_i^0)}{T}}{\partial \frac{1}{T}}\right]_{P, n_i} = (\bar{H}_i - H_i^0), \tag{18.35}$$

$$\left(\frac{\partial \bar{G}_i}{\partial P}\right)_{T, n_i} = \bar{V}_i$$

$$\left[\frac{\partial(\bar{G}_i - G_i^0)}{\partial P}\right]_{T, n_i} = (\bar{V}_i - V_i^0), \tag{18.36}$$

in which S_i^0, H_i^0, V_i^0 are the appropriate molar properties of i in the pure state.

CALCULATION OF PARTIAL PROPERTIES FROM TOTAL PROPERTIES

It is also possible, given the total molar properties of a system as a function of composition, to obtain values for the corresponding partial molar properties at any given composition. We will develop these relations for the case of the total and partial molar Gibbs free energies (the chemical potentials) in a two-component system at constant temperature and pressure. We will assume that we have available the data required to plot G as a function of composition. Such a plot is shown in Figure 18.1. Let us consider the properties of this system at some particular composition, say a in Figure 18.1. For a small composition change about point a, at equilibrium at constant temperature and pressure, we may write

$$dG = \bar{G}_A dX_A + \bar{G}_B dX_B. \tag{18.37}$$

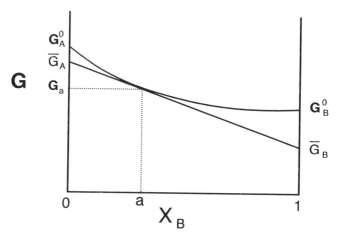

Figure 18.1 Gibbs free energy versus composition diagram for a one-phase, two-component system, showing the method for obtaining the partial molar Gibbs free energies, $\bar{G}_i$, from the **G** versus X_B curve.

Moreover, since we know that

$$X_A + X_B = 1, \qquad dX_A = -dX_B, \tag{18.38}$$

we may write, in order to eliminate dX_A,

$$d\boldsymbol{G} = (\bar{G}_B - \bar{G}_A)dX_B, \tag{18.39}$$

or

$$\bar{G}_B = \bar{G}_A + \left(\frac{d\boldsymbol{G}}{dX_B}\right). \tag{18.40}$$

We can eliminate $\bar{G}_A$ from the above expression by using the integrated expression

$$\boldsymbol{G} = \bar{G}_A X_A + \bar{G}_B X_B = \bar{G}_A(1 - X_B) + \bar{G}_B X_B, \tag{18.41}$$

or

$$\bar{G}_A = \frac{\boldsymbol{G} - \bar{G}_B X_B}{1 - X_B}. \tag{18.42}$$

Substituting this expression into the above expression for $\bar{G}_B$ yields

$$\bar{G}_B = \frac{(\boldsymbol{G} - \bar{G}_B X_B)}{(1 - X_B)} + \left(\frac{d\boldsymbol{G}}{dX_B}\right) \tag{18.43}$$

or

$$\bar{G}_B(1 - X_B) + \bar{G}_B X_B = G + (1 - X_B)\left(\frac{dG}{dX_B}\right), \quad (18.44)$$

or

$$\bar{G}_B = G + (1 - X_B)\left(\frac{dG}{dX_B}\right). \quad (18.45)$$

By a similar argument we can show that

$$\bar{G}_A = G - X_B\left(\frac{dG}{dX_B}\right). \quad (18.46)$$

These results can be shown geometrically in Figure 18.1 by drawing the tangent to the G versus composition curve at composition a. It can be seen that $\bar{G}_A$ is given by the intercept of this line with the $X_B = 0$ axis, while $\bar{G}_B$ is given by the intersection with the $X_B = 1$ axis.

SUMMARY

We have now developed all of the concepts that we will need in the study of multicomponent condensed phases, or solutions. In this study we will use the definition of activity,

$$\mu_i = \mu_i^0 + kT \ln a_i, \quad (18.47)$$

the various total molar properties,

$$G, \ H, \ S, \ E, \ V, \quad (18.48)$$

and the partial molar properties

$$\bar{G}_i, \ \bar{H}_i, \ \bar{S}_i, \ \bar{E}_i, \ \bar{V}_i. \quad (18.49)$$

We will also use the relations between the partial and total molar properties, such as

$$G = \sum_i X_i \bar{G}_i, \quad V = \sum_i X_i \bar{V}_i, \quad (18.50)$$

the relationships among the partial molar properties, such as

$$\bar{G}_i = \bar{H}_i - T\bar{S}_i, \quad (18.51)$$

CLASSICAL THERMODYNAMICS OF MULTICOMPONENT SYSTEMS

and the relations that allow evaluation of the partial molar properties in terms of the total properties, such as

$$\bar{G}_B = G + (1 - X_B)\left(\frac{dG}{dX_B}\right),$$

$$\bar{G}_A = G - X_B\left(\frac{dG}{dX_B}\right). \qquad (18.52)$$

These are all the tools that we will need to treat solutions on a classical thermodynamic basis.

BIBLIOGRAPHY

R. T. DeHoff, *Thermodynamics in Materials Science*, McGraw-Hill, New York, 1993, Chapter 8.

D. R. Gaskill, *Introduction to Metallurgical Thermodynamics*, McGraw-Hill, New York, 1973, Chapter 11.

E. A. Guggenheim, *Thermodynamics*, North Holland, 1957, Chapter 5.

D. V. Ragone, *Thermodynamics of Materials*, Volume 1, John Wiley & Sons, New York, 1995, Chapter 7.

R. A. Swalin, *Thermodynamics of Solids*, John Wiley & Sons, New York, 1962, Chapter 8.

PROBLEMS

18.1 Beginning with the general relation

$$F = E - TS,$$

develop the relation among the partial molar properties that

$$(\bar{F}_i - F_i^0) = (\bar{E}_i - E_i^0) - T(\bar{S}_i - S_i^0).$$

18.2 Develop relations between the partial molar volumes in a binary system, $\bar{V}_A$ and $\bar{V}_B$, and the total molar volume, V, as a function of composition, similar to the relations developed in the text for the partial molar Gibbs free energies.

18.3 Suppose the vapor pressure of a specific concentration of zinc dissolved in copper is 3×10^{-3} mm Hg (0.00133 Pa) at 600°C. Calculate the activity of zinc in the alloy. Calculate the free energy change of zinc upon solution—that is, $(\bar{G}_{Zn} - G_{Zn}^0)$.

18.4 The activity of nickel in iron at a concentration of $X_{Ni} = 0.1$ and a temperature of 1600°C is 0.067. Calculate the equilibrium partial pressure of nickel over the alloy.

18.5 The molar Gibbs free energy of a binary system at a particular temperature and pressure is given by

$$G = 200X_A + 150X_B - 50X_AX_B \text{ J/mol.}$$

Calculate the partial molar Gibbs free energies for A and B at a composition of $X_B = 0.3$.

CHAPTER 19

CLASSICAL THERMODYNAMICS OF SOLUTIONS

Let us now go on to consider the changes in the thermodynamic properties of the system that take place when two pure components are mixed to form a one-phase mixture, be it solid, liquid, or gas. Such mixtures are in general referred to as *solutions*. The general thermodynamic framework that we will develop applies equally well to all cases, but the detailed results will, of course, depend on the atomic or molecular species involved as well as variables such as pressure and temperature.

FORMATION OF A SOLUTION

In general, we will be looking at a reaction of the form

$$X_A \text{ mol of } A + X_B \text{ mol of } B \rightarrow 1 \text{ mol of solution of composition } A_{X_A} \cdot B_{X_B}. \tag{19.1}$$

The Gibbs free energy change associated with this reaction, per mole, is

$$\Delta G_M = G^{soln} - X_A G_A^0 - X_B G_B^0, \tag{19.2}$$

or, in terms of chemical potentials,

$$\Delta G_M = G^{soln} - X_A N_{av} \mu_A^0 - X_B N_{av} \mu_B^0. \tag{19.3}$$

Since we know that

$$G^{soln} = X_A \bar{G}_A + X_B \bar{G}_B, \qquad (19.4)$$

we may write

$$\Delta G_M = X_A(\bar{G}_A - G_A^0) + X_B(\bar{G}_B - G_B^0) \qquad (19.5)$$

for this case of a two-component system. For the general case of a multicomponent system we have

$$\Delta G_M = \sum_i X_i(\bar{G}_i - G_i^0). \qquad (19.6)$$

Alternatively, we may substitute for $(\bar{G}_i - G_i^0)$ in terms of the activity through the expression

$$(\bar{G}_i - G_i^0) = N_{av}(\mu_i - \mu_i^0) = RT \ln a_i, \qquad (19.7)$$

yielding

$$\Delta G_M = RT(X_A \ln a_A + X_B \ln a_B) \qquad (19.8)$$

or, in general

$$\Delta G_M = RT \sum_i X_i \ln a_i. \qquad (19.9)$$

Similarly, we may write

$$\Delta H_M = \sum_i X_i(\bar{H}_i - H_i^0)$$

$$\Delta S_M = \sum_i X_i(\bar{S}_i - S_i^0) \qquad (19.10)$$

$$\Delta V_M = \sum_i X_i(\bar{V}_i - V_i^0).$$

These relations are quite general, and they apply to all single-phase multicomponent systems.

IDEAL GAS MIXTURES

The application of the relations developed above to a mixture of ideal gases is straightforward. Such mixtures obey Dalton's law of partial pressures, namely that

$$P = \sum_i P_i, \qquad (19.11)$$

where the P_i are the partial pressures of the individual gases prior to the mixing process. We may substitute for P_i that

$$P_i = \frac{n_i RT}{V}, \qquad (19.12)$$

leading to

$$P = \sum_i P_i = \sum_i \frac{n_i RT}{V} \qquad (19.13)$$

and consequently

$$X_i \equiv \frac{n_i}{\sum_i n_i} = \frac{P_i}{\sum_i P_i}. \qquad (19.14)$$

Furthermore, we know that for an ideal gas

$$a_i = \frac{P_i}{P_i^0}, \qquad (19.15)$$

where the choice of P_i^0 is arbitrary. In the present case, it is convenient to choose P_i^0 as

$$P_i^0 = \sum_i P_i, \qquad (19.16)$$

leading to

$$a_i = \frac{P_i}{\sum_i P_i} = X_i. \qquad (19.17)$$

This result may be substituted into the general expression for ΔG_M [equation (19.9)] to yield

$$\Delta G_M = RT \sum_i X_i \ln X_i, \qquad (19.18)$$

or, in terms of pressures,

$$\Delta G_M = RT \sum_i \left(\frac{P_i}{\sum_i P_i}\right) \ln\left(\frac{P_i}{\sum_i P_i}\right). \tag{19.19}$$

In order to obtain the changes in the other thermodynamic properties associated with mixing, we may use the relation for a_i above and write that

$$(\bar{G}_i - G_i^0) = N_{av}(\mu_i - \mu_i^0) = RT \ln a_i = RT \ln X_i \tag{19.20}$$

for this case. This relation may be used, along with the general relation

$$\left(\frac{\partial \frac{\bar{G}_i - G_i^0}{T}}{\partial \frac{1}{T}}\right)_{P,n_i} = (\bar{H}_i - H_i^0), \tag{19.21}$$

to yield

$$(\bar{H}_i - H_i^0) = \left(\frac{\partial \frac{RT \ln X_i}{T}}{\partial \frac{1}{T}}\right)_{P,n_i} = \frac{\partial}{\partial \frac{1}{T}}(R \ln X_i) = 0, \tag{19.22}$$

as $X_i \neq f(T)$. Thus

$$\Delta H_M = \sum_i (\bar{H}_i - H_i^0) = 0 \tag{19.23}$$

for a mixture of ideal gases. Finally, using

$$\Delta G_M = \Delta H_M - T\Delta S_m, \tag{19.24}$$

we have, since $\Delta H_M = 0$,

$$\Delta S_M = -R \sum_i X_i \ln X_i. \tag{19.25}$$

We will see later that this is the expression for the ideal entropy of mixing which can also be derived from statistical considerations.

MULTICOMPONENT CONDENSED PHASES

Let us look now at multicomponent single-phase condensed systems, or, as they are usually called, *solutions*. The activity of a given component in a solution can be related to the equilibrium partial pressure of that component in the vapor phase in equilibrium with the solution as follows: In the vapor phase, if we assume the vapor phase to be an ideal gas mixture, we have

$$\mu_i^V = (\mu_i^0)^V + kT \ln \frac{P_i}{P_i^0}. \tag{19.26}$$

In the condensed phase we have

$$\mu_i^C = (\mu_i^0)^C + kT \ln a_i. \tag{19.27}$$

Since the two phases are in equilibrium, we obtain

$$\mu_i^V = \mu_i^C. \tag{19.28}$$

Since in the standard state the pure vapor and pure condensed phase are in equilibrium, we have

$$(\mu_i^0)^V = (\mu_i^0)^C. \tag{19.29}$$

Thus

$$kT \ln \frac{P_i}{P_i^0} = kT \ln a_i, \tag{19.30}$$

or

$$a_i = \frac{P_i}{P_i^0}. \tag{19.31}$$

This is a general relation for any component in a solution whose vapor phase is an ideal gas.

With this result in mind, let us look at the values that P_i can assume over the full range of solution compositions. For example, for a two-component solution we know that, for pure A at $X_A = 1$, $P_A = P_A^0$, because this corresponds to the pure component A, and similarly at $X_A = 0$, $P_A = 0$, because there is no A in the system. Similar expressions can be written for X_B at the concentration extremes of the system. These relations can be shown on a plot of P_i versus X_B for the system, as shown in Figure 19.1.

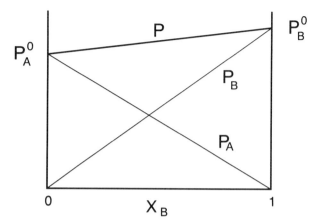

Figure 19.1 Partial pressures of the two components and system total pressure as a function of composition for a system obeying Raoult's law (an ideal system).

THE IDEAL SOLUTION

The next question is that of how P_A and P_B change with composition between the limits deduced above. The way in which they vary will depend on the properties of the solution. The simplest possible dependence is a linear relation between P_i and X_i, namely,

$$P_i = X_i P_i^0. \tag{19.32}$$

Solutions that obey this relation, which is known as *Raoult's law*, are called *ideal solutions*. The behavior of P_A and P_B over the full range of X_B is shown in Figure 19.1. For this case, the activity of a given component in the solution is simply

$$a_i = \frac{P_i}{P_i^0} = \frac{X_i P_i^0}{P_i^0} = X_i, \tag{19.33}$$

and the curves of P_A and P_B versus X_B in Figure 19.1 are straight lines. Note that this is the same result that we obtained for the case of the ideal gas mixture. As a consequence of this, all of the changes in the thermodynamic properties associated with the formation of an ideal solution will be identical to those already deduced for an ideal gas mixture, namely,

$$\Delta G_M = RT \sum_i X_i \ln X_i,$$

$$\Delta H_M = 0, \tag{19.34}$$

$$\Delta S_M = -R \sum_i X_i \ln X_i.$$

In addition, we may determine the volume change on mixing using the general expression

$$(\bar{V}_i - V_i^0) = \left(\frac{\partial(\bar{G}_i - G_i^0)}{\partial P}\right)_{T,n_i} = \left(\frac{\partial(RT \ln X_i)}{\partial P}\right)_{T,n_i} = 0. \quad (19.35)$$

In this case, since $X_i \neq f(P)$, we have

$$\Delta V_M = \sum_i X_i(\bar{V}_i - V_i^0) = 0. \quad (19.36)$$

Thus we see that the ideal solution is the exact analog of the ideal gas mixture. It is made up of essentially noninteracting particles. A solution made up of red and green marbles would have the same thermodynamic mixing properties as our ideal solution in which mixing takes place on a molecular level.

DILUTE SOLUTIONS

Now let us go on to consider more complicated systems in which the presence of one component *does* have an effect on the properties of the other components, and vice versa. This case cannot be treated simply, although we can make some generalizations about certain special cases or certain limited ranges of composition. We will treat some of these special cases later by statistical mechanics. For the present, we will confine ourselves to discussing the behavior of a few classes of solutions in general terms.

Consider first what are known as *dilute solutions*. The dilute solution is defined as one in which the concentration of the solute is sufficiently low that solute atoms do not "see" one another. That is, each solute atom acts as though it were in an environment of pure solvent. As a result, the thermodynamic properties of the solute vary linearly with composition. Consequently, the equilibrium vapor pressure of the solute (call it A) over the solution, assuming the vapor phase to be ideal, will be

$$P_A = bX_A. \quad (19.37)$$

This is known as *Henry's law*, and it is actually observed in practice for many systems over a limited but finite composition range. In this case, we may write the following equation for the activity of A:

$$a_A = \frac{P_A}{P_A^0} = \frac{bX_a}{P_A^0} = \gamma_A^0 X_A, \quad (19.38)$$

where $\gamma_A^0 \equiv b/P_A^0$ is defined as activity coefficient of A at infinite dilution. Henry's law assumes that γ^0 is constant over the range of applicability of the law. For this case, if we plot a_i versus X_B, for the case of a two-component solution, we again have

$$\begin{array}{llll} \text{For pure } A: & X_A = 1, & P_A = P_A^0, & a_A = 1, \\ \text{For pure } B: & X_A = 0, & P_A = 0, & a_A = 0, \end{array} \quad (19.39)$$

and similarly for component B. In this case, however, at finite values of X_A we do not have

$$a_A = X_A, \quad (19.40)$$

but, for low concentrations of A in B, we have

$$a_A = \gamma_A^0 X_A, \quad (19.41)$$

where γ_A^0 may be either greater than or less than unity, depending on the nature of the interaction between A atoms and B atoms. Similar relations hold in the vicinity of $X_A = 1$. This behavior is shown schematically in Figure 19.2.

We may next consider the question of the behavior of the solvent in a dilute solution. We have previously shown that, at constant temperature and pressure, for a small composition change about a point of equilibrium, we have, in general,

$$X_A d\bar{G}_A = X_B d\bar{G}_B = 0, \quad (19.42)$$

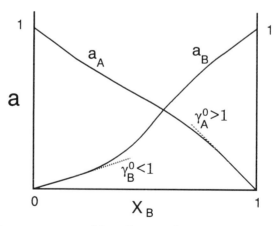

Figure 19.2 Activity versus composition diagram for a nonideal system. At low concentrations of A or B, the solute obeys Henry's law, with a constant $\gamma_i^0 \neq 1$, while the solvent obeys Raoult's law with $\gamma_i^0 = 1$.

or, recalling that

$$d\bar{G}_i = RT d\ln a_i, \tag{19.43}$$

we have

$$X_A d\ln a_A + X_B d\ln a_B = 0. \tag{19.44}$$

For the case of a dilute solution of B in A,

$$a_B = \gamma_B^0 X_B, \tag{19.45}$$

with γ_B^0 a constant. Thus we may write

$$X_A d\ln a_A + X_B d\ln X_B + X_B d\ln \gamma_B^0 = 0, \tag{19.46}$$

or, since γ_B^0 is a constant,

$$X_A d\ln a_A = -X_B d\ln X_B, \tag{19.47}$$

or

$$d\ln a_A = -\left(\frac{X_B}{X_A}\right) d\ln X_B$$

$$= -\left(\frac{X_B}{X_A}\right)\left(\frac{1}{X_B}\right) dX_B$$

$$= -\frac{dX_B}{X_A}. \tag{19.48}$$

Or, since $dX_A = -dX_B$, we have

$$d\ln a_A = \frac{dX_A}{X_A} = d\ln X_A, \tag{19.49}$$

or, by integrating between the appropriate limits we obtain

$$\int_1^{a_A} d\ln a_A = \int_1^{X_A} d\ln X_a, \tag{19.50}$$

$$a_A = X_A,$$

which is simply Raoult's law. Thus for the range of composition over which the solute obeys Henry's law, the solvent obeys Raoult's law. This behavior is also shown in Figure 19.2. The range of composition over which dilute solution behavior is observed will vary from system to system. In general, the closer γ_i^0 is to unity, the greater the composition range over which dilute solution behavior is observed.

CONCENTRATED SOLUTIONS

In the composition range beyond that where dilute solution behavior is observed, no easy generalizations can be made. In general, the activity in this case, known as the *concentrated solution case*, may be represented by

$$a_i = \gamma_i X_i, \tag{19.51}$$

in which γ_i, the activity coefficient, is some function of the composition. We will not consider this case in detail.

We will, however, define one other class of solution behavior, because we will see it again when we consider the statistical mechanical treatment of solutions. This is the so-called regular solution. This is a solution in which the enthalpy of mixing has a finite value, but the entropy of mixing has the ideal value given by equation (19.34), namely,

$$\Delta S_M = -R(X_A \ln X_A + X_B \ln X_B). \tag{19.52}$$

We shall find that this definition, which approximates the behavior of many real systems, allows us to treat the behavior of solutions on a statistical basis without encountering unduly difficult mathematical problems.

EXCESS FUNCTIONS

As a final point on the classical treatment of multicomponent systems, we will find that it is often convenient to describe the changes in the properties occurring on mixing in terms of the departure from ideal solution behavior. To do this, we define so-called *excess* thermodynamic properties as follows: For a nonideal solution, in general, we may write the following equation for the change in chemical potential of a given component with mixing:

$$\Delta \mu_i = (\mu_i - \mu_i^0) = kT \ln a_i = kT \ln \gamma_i + kT \ln X_i. \tag{19.53}$$

For an ideal solution, $\gamma_i = 1$ and

$$(\Delta \mu_i)^{ideal} = kT \ln X_i. \tag{19.54}$$

We may use this result to define

$$(\Delta \mu_i)^{xs} = (\Delta \mu_i) - (\Delta \mu_i)^{ideal}$$
$$= kT \ln \gamma_i, \qquad (19.55)$$

or, in terms of partial molar properties

$$(\Delta \bar{G}_i)^{xs} = RT \ln \gamma_i. \qquad (19.56)$$

We may apply this result to the previously developed expressions for the changes in the thermodynamic properties associated with mixing to obtain

$$(\Delta G_M)^{xs} = \Delta G_M - (\Delta G_M)^{ideal}$$
$$= RT(X_A \ln X_A + X_A \ln \gamma_A + X_B \ln X_B + X_B \ln \gamma_B)$$
$$- RT(X_A \ln X_A + X_B \ln X_B)$$
$$= RT(X_a \ln \gamma_A + X_B \ln \gamma_B). \qquad (19.57)$$

Similarly,

$$(\Delta H_M)^{xs} = \Delta H_M - (\Delta H_M)^{ideal}$$
$$= \Delta H_M, \qquad (19.58)$$

or

$$(\Delta \bar{H}_i)^{xs} = \Delta \bar{H}_i; \qquad (19.59)$$

and finally

$$(\Delta V_M)^{xs} = \Delta V_M - (\Delta V_M)^{ideal}$$
$$= \Delta V_M, \qquad (19.60)$$

or

$$(\Delta \bar{V}_i)^{xs} = \Delta \bar{V}_i. \qquad (19.61)$$

Note that because the ideal solution is defined as one in which no net interactions occur, the excess quantities defined above must be related to the extent of interaction that does occur. We will see this more clearly when we treat the properties of solutions statistically, using what are called the *Bragg–Williams* and the *quasichemical* approaches.

BIBLIOGRAPHY

R. T. DeHoff, *Thermodynamics in Materials Science*, McGraw-Hill, New York, 1993, Chapter 8.

D. R. Gaskill, *Introduction to Metallurgical Thermodynamics*, McGraw-Hill, New York, 1973, Chapter 11.

P. Gordon, *Principles of Phase Diagrams in Materials Systems*, McGraw-Hill, New York, 1968, Chapter 4.

E. A. Guggenheim, *Thermodynamics*, North Holland, 1957, Chapter 6.

H. Reiss, *Methods of Thermodynamics*, Blaisdell, New York, 1965, Chapter 9.

PROBLEMS

19.1 Derive the expressions for $(\bar{H}_B - H_B^0)$ and $(\bar{V}_B - V_B^0)$ for the solute in a dilute solution.

19.2 Solid Au–Cu alloys are regular in their thermodynamic behavior. ΔH_M at 773 K is shown in the table below.
  (a) Find $\Delta \bar{H}_{Cu}$ and $\Delta \bar{H}_{Au}$ at $X_{Cu} = 0.3$.
  (b) Find ΔG_M at $X_{Cu} = 0.3$.
  (c) Calculate the equilibrium partial pressure of Cu over an alloy containing 30 At% Cu and 70 At% Au.

X_{Cu}:	0.1	0.2	0.3	0.4	0.5	0.6	0.7	0.8	0.9
ΔH_M:	−355	−655	−910	−1120	−1230	−1240	−1130	−860	−460

19.3 Components A and B form a regular solution in the solid state for which $\Delta H_M = 20{,}000\, X_A \cdot X_B$ J/mol. What is the equilibrium solubility of B in A at 1000 K?

19.4 For the system described in Problem 19.3, plot the chemical potentials of A and B atoms as a function of composition at 1200 K.

19.5 At a particular temperature, dilute solutions of A in B have been observed to follow Henry's law, with $\gamma_A^0 = 0.9$. If $P_A^0 = 1 \times 10^2$ Pa and $P_B^0 = 3 \times 10^3$ Pa, calculate the equilibrium partial pressures of A and B over the alloy.

CHAPTER 20

LATTICE STATISTICS

As a first step in developing the methods required to treat multicomponent, condensed-phase systems, we will develop the method known as *lattice statistics* and apply it to the case of the binding of noninteracting atoms or molecules to a lattice of sites. A system of particles bound to such a lattice of sites is called a *lattice gas*. For the case considered here, where the particles bound to the lattice do not interact with each other, we may refer to the system as an *ideal lattice gas*.

THE IDEAL LATTICE GAS

The model that we will deal with here assumes that there exists a regular array of binding sites, in one, two, or three dimensions. The atoms or molecules whose thermodynamic behavior we will deduce are bound to these sites, no more than one per site. These bound atoms or molecules are assumed to vibrate about their equilibrium positions in three dimensions. A typical one-dimensional system of this type would be molecules bound to a linear polymer chain. A two-dimensional system would be the adsorption of molecules onto the surface of a crystalline solid. The simplest three-dimensional model would be an Einstein crystal with, in general, allowance for vacant lattice sites. In all of these cases, we have a number, N, of particles bound to a number, M, of sites, where in all cases $N \leqslant M$. These are not, strictly speaking, multicomponent systems. We shall see, however, that when we do begin explicit consideration of multicomponent systems, the roles played in the present case by filled and empty sites will be played in the multicomponent system by sites filled by two different components.

We will develop expressions for the partition function, and from these the thermodynamic functions, for two examples of ideal lattice gases. We will consider one three-dimensional case, namely the Einstein crystal with vacant lattice sites, and one two-dimensional case, the adsorption of molecules from a gas phase onto a crystalline solid surface, the so-called Langmuir model of gas adsorption.

THE EINSTEIN CRYSTAL WITH VACANCIES

Consider first the Einstein crystal with vacancies. Recall that in our previous treatment of the Einstein solid, we used a model in which atoms were held at lattice sites by the potential well arising from the interaction of each atom with all of its neighbors, the interaction energy being u_r. The atoms were assumed to vibrate about their equilibrium positions in three independent one-dimensional harmonic vibrational modes, all having the same frequency. These assumptions led to an expression for q for each atom of the form

$$q = q_0 \cdot q_v^3, \tag{20.1}$$

$$q_0 = e^{-u_0/2kT}, \qquad q_v = e^{-h\nu/2kT} \sum_n e^{-nh\nu/kT}. \tag{20.2}$$

Because we were dealing with a system of independent, identical, distinguishable particles, the expression for the partition function, Q, was

$$Q = q^N = q_0^N \cdot q_v^{3N} = e^{-Nu_0/2kT} \cdot q_v^{3N}. \tag{20.3}$$

In the original treatment of the Einstein solid in Chapter 14, we assumed that the number of atoms in the crystal, N, was equal to the number of lattice sites. In other words, vacancies were forbidden. We now relax this restriction, so that the total number of lattice sites, M, can be greater than N. This introduces additional microstates into the sum leading to Q, because each possible way of arranging the N atoms on the M distinguishable lattice sites represents a different microstate. We must therefore correct the expression for Q shown in equation (20.3) by multiplying it by the expression for the number of ways of arranging N identical particles on M distinguishable sites with no more than one particle per site. The required expression, which is found in any standard text on statistics, is

$$\frac{M!}{N!(M-N)!}. \tag{20.4}$$

We must also allow for the fact that introducing the vacant lattice sites reduces the total number of atom–atom interactions per unit volume in the system and consequently reduces the electronic contribution to Q, because there are now some of the M sites that are unoccupied and thus do not contribute to the binding energy of the crystal. We will assume that the extent of this reduction is $u_0/2$ for each vacancy added. In reality, this is an overestimate, because it neglects any relaxation of the atoms adjacent to the vacancy, but it will serve to illustrate the effect of added vacancies on the thermodynamic properties of the system. Making this change leads to

$$Q = e^{-(Nu_0/2kT) + (M-N)(u_0/2kT)} q_v^{3N} = e^{(M-2N)(u_0/2kT)} q_v^{3N}. \quad (20.5)$$

The final expression for Q is thus

$$Q(N, M, T) = \frac{M!}{N!(M-N)!} e^{(M-2N)(u_0/2kT)} q_v^{3N}. \quad (20.6)$$

Using Stirling's approximation on the factorial terms, we obtain

$$\ln Q = M \ln M - N \ln N - (M - N) \ln(M - N)$$
$$+ (M - 2N) \left(\frac{u_0}{2kT}\right) + 3N \ln q_v. \quad (20.7)$$

EVALUATION OF THE THERMODYNAMIC FUNCTIONS

We are now faced with the usual problem of obtaining expressions for the thermodynamic functions in terms of Q. In this case, we have a complicating factor, in that Q is a function of the variables N, M, and T rather than N, V, and T as in previous cases. This is more a complication in appearance than in fact, because a system having a fixed number of sites per unit dimension (length, area, or volume) is formally the same problem as treated in our development of the canonical ensemble. That is, we have a prototype system that can exchange energy with its surroundings, but not material; and, formally, a fixed M is equivalent to a fixed V. We can thus use equation (20.8) as our basic equation relating Q to the classical thermodynamic state functions:

$$F = -kT \ln Q(N, M, T), \quad (20.8)$$

where for this case

$$F = -kT \bigg[M \ln M - N \ln N - (M - N) \ln(M - N)$$
$$+ (M - 2N) \left(\frac{u_0}{2kT}\right) + 3N \ln q_v \bigg]. \quad (20.9)$$

This expression differs from that obtained for the Einstein solid by the additive factor

$$kT\left[M \ln M - N \ln N - (M-N)\ln(M-N) + (N-M)\left(\frac{u_0}{2kT}\right)\right]. \quad (20.10)$$

The expression for energy is

$$E = kT^2 \left(\frac{\partial \ln Q}{\partial T}\right)_{N,M}$$

$$= kT^2 \frac{\partial}{\partial T}\left[M \ln M - N \ln N - (M-N)\ln(M-N)\right.$$

$$\left. + (M-2N)\left(\frac{u_0}{2kT}\right) + 3N \ln q_v\right]_{M,N}$$

$$= (M-2N)\frac{u_0}{2} + 3NkT^2 \left(\frac{\partial \ln q_v}{\partial T}\right)_{M,N}. \quad (20.11)$$

That is, the value of E, for a given N, is less negative than that for the Einstein crystal due to the reduction in the total binding energy associated with the interactions lost in the formation of the $M - N$ vacancies.

C_V will be correspondingly changed, because it is given, as usual, by

$$\left(\frac{\partial E}{\partial T}\right)_{M,N},$$

and the temperature dependence of E includes the temperature dependence of the vacancy concentration. This is a small correction and will not be considered in detail here.

The entropy is calculated from

$$S = \frac{E-F}{T}$$

$$= (2N-M)\frac{u_0}{2T} + 3NkT\left(\frac{\partial \ln q_v}{\partial T}\right)_{M,N} - (2N-M)\frac{u_0}{2T} + 3Nk \ln q_v$$

$$+ k[M \ln M - N \ln N - (M-N)\ln(M-N)]$$

$$= k[M \ln M - N \ln N - (M-N)\ln(M-N)] + 3Nk\left[T\left(\frac{\partial \ln q_v}{\partial T}\right)_{M,N} + \ln q_v\right], \quad (20.12)$$

or
$$S = S_{config} + S_{vib}, \tag{20.13}$$

where in the last relation we have separated the contribution due to the possible ways of arranging the N atoms on the M sites (the configurational entropy) from the contributions due to the vibrations of the atoms about their equilibrium positions in the lattice (the vibrational entropy). This separation is a quite common one in systems of more than one component, or where there are filled and empty sites.

It is of interest to make one further manipulation with the thermodynamic expressions for the Einstein crystal with vacancies—namely, to develop an expression for the equilibrium vacancy concentration. Recalling the discussion of criteria for equilibrium presented in Chapter 3, we know that for this system of constant M and T, which we have shown to be formally equivalent to a system of constant V and T, the criterion for equilibrium is a minimum in F. Since we can write F as a function of $Q(N, M, T)$, the desired minimum will be that at which

$$\left(\frac{\partial F}{\partial M}\right)_{N,T} = 0, \tag{20.14}$$

or, expressing F in terms of Q,

$$\frac{\partial}{\partial M}[-kT \ln Q(N, M, T)]_{N,T} = 0. \tag{20.15}$$

Substituting for $\ln Q$ and performing the differentiation yields

$$-kT\frac{\partial}{\partial M}\left[M \ln M - N \ln N - (M-N) \ln(M-N) + (M-2N)\frac{u_0}{2kT} + 3N \ln q_v\right] = 0$$

$$-kT\frac{\partial}{\partial M}\left[M \ln M - (M-N) \ln(M-N) + \frac{Mu_0}{2kT}\right] = 0$$

$$-kT\left[\ln M - \ln(M-N) + \frac{u_0}{2kT}\right] = 0. \tag{20.16}$$

Since kT is always finite and positive, the only way this equation can be satisfied is if

$$\ln \frac{M-N}{M} = \frac{u_0}{2kT}. \tag{20.17}$$

We can verify that this extremum in F is a minimum by taking the second derivative

$$\left(\frac{\partial^2 F}{\partial M^2}\right) = \frac{\partial}{\partial M}\left[kT\left(\ln\frac{M-N}{M} - \frac{u_0}{2kT}\right)\right]$$

$$= kT\left[\frac{1}{(M-N)} - \frac{1}{M}\right]. \tag{20.18}$$

Since $M - N$ will always be smaller than M, this second derivative is positive and the extremum calculated is a minimum. The equilibrium fraction of vacant sites is thus

$$\frac{M-N}{M} = e^{u_0/2kT}, \tag{20.19}$$

in agreement with the observation that the equilibrium vacancy concentration increases exponentially with temperature, with the rate of increase related to the binding energy of the crystal.

THE LANGMUIR MODEL OF ADSORPTION

Let us consider now a second case of an ideal lattice gas, the Langmuir model of adsorption. Let us begin by considering first the nature of the binding forces between an atom from the gas phase and the crystal surface. The surface presents a regular, two-dimensional array of atoms, as shown in Figure 20.1. The atom or molecule in the gas phase will interact with a large number of surface and near-surface atoms, the strength of the interaction varying both with the distance of the atom from the surface, shown as the z direction in Figure 20.1, and with the lateral position of the atom relative to the surface atoms, shown as the x and y directions in Figure 20.1. The total interaction energy can be given as the sum of the interactions between the gas phase particle and all of the atoms in the solid. Formally, this can be written as

$$u(x, y, z) = \sum_i u(r_i) = F(x, y, z), \tag{20.20}$$

where the $u(r_i)$ represent the pairwise interactions between the gas-phase species and each surface atom, and the sum is taken over all surface and near-surface atoms that are close enough to the gas-phase species to make a significant contribution. The nature of this interaction energy can be shown schematically either by looking at $u(x, y, z)$ as a function of z at a given x, y or by looking at the variation of $u(x, y, z)$ as a function of x or y at some equilibrium value of z. The form of $u(z)$ versus z is shown schematically in

THE LANGMUIR MODEL OF ADSORPTION   241

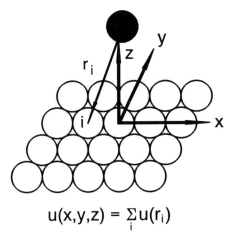

Figure 20.1 The model used in calculating the potential energy of an adsorbed species as a function of position relative to a surface by summing pairwise interactions.

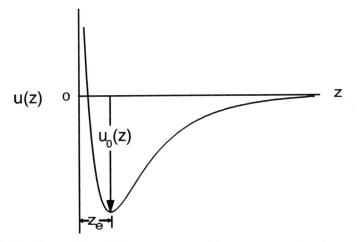

Figure 20.2 The variation of the system potential energy as an atom from a gas phase is moved normal to a surface at constant x, y position relative to the surface lattice. The pairwise interactions involved are assumed to be of the Lennard-Jones 6-12 form.

Figure 20.2. Here the zero of potential energy is taken as the gas-phase species at infinite distance from the surface, to be consistent with our previous definition of the zero of energy. As z decreases, the energy of the system becomes negative, implying attractive forces between the gas-phase species and the surface, reaches maximum negative value at some value z_e, then rises and becomes positive (repulsive) at very small values of z. The overall shape of the curve is similar to those developed for other systems that we have treated using pairwise interaction models. This is to be expected, as the curve in the present

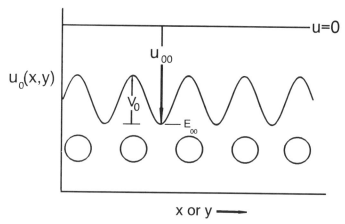

Figure 20.3 The variation of the minimum in the adsorbed atom potential well depth, u_0, as an adsorbed atom is moved parallel to the surface of a crystal.

case also arose from summing pairwise interactions between the adsorbed species and the atoms in the solid.

The variation of $u(x, y, z)$ as a function of x or y is most conveniently expressed in terms of the variation of the minimum of the $u(z)$ curve, u_0, with x or y. A schematic plot of u_0 versus x or y is shown in Figure 20.3. The regular variation of u_0 with x or y reflects the regular interatomic spacing on the crystal surface. This plot indicates that there are favored values of x (and y) at which u_0 has a maximum negative value. These sites, characterized by u_{00} on the figure, represent the equilibrium binding sites for the species adsorbed on the surface. The intervening minima in the absolute value of u_0, designated by v_0, represent less stable configurations that the adsorbed species must pass through in order to migrate from one favored site to another. If the value of kT is small compared to v_0, negligible migration will occur, and the adsorbed species can be considered as bound at the equilibrium sites.

The assumptions of the Langmuir model for adsorption on a solid surface are that the adsorbed species are bound to adsorption sites, that they cannot migrate from site to site, and that they vibrate about the equilibrium position in three dimensions. The vibrational motion of one adsorbed particle is taken to be independent of the other adsorbed particles present, and the three vibrational modes of a given adsorbed particle are assumed independent of one another. This leads to an expression for q for each molecule of

$$q = q_x \cdot q_y \cdot q_z \cdot q_0. \tag{20.21}$$

The term q_0 represents the binding energy of the adsorbed species, and can be given by

$$q_0 = e^{-u_{00}/kT}, \tag{20.22}$$

analogously to the previously studied cases of the Einstein and Debye solids. Here u_{00} represents the potential at the bottom of the well. As in the case of the crystalline solids, u_{00} is inherently negative and accounts for the difference in the system energy relative to our previously chosen zero of energy due to the binding forces holding the adsorbed species to the surface. The terms q_x, q_y, and q_z are one-dimensional harmonic quantum oscillator partition functions similar to those we have used previously. In the present case, since the potential well in which the adsorbed species vibrates is not spherically symmetric, all three modes will in general have different frequencies. Note that the lowest energy allowed to an adsorbed species is

$$e_{ads} = u_{00} + \left(\frac{h}{2}\right)(v_x + v_y + v_z). \tag{20.23}$$

This parameter, known as the *energy of adsorption*, can be measured experimentally by a variety of techniques.

The considerations involved in going from q to Q in this case are similar to those for the Einstein crystal with vacancies. The contribution to Q from the allowed energies of the adsorbed species, which for this case of identical, distinguishable adsorbed atoms is q^N, must be multiplied by the number of ways of arranging the N adatoms on the M lattice sites as before to yield

$$Q(N, M, T) = \frac{M! q^N}{N!(M - N)!}, \tag{20.24}$$

or, using Stirling's approximation on the factorial terms,

$$\ln Q = M \ln M - N \ln N - (M - N) \ln(M - N) + N \ln q. \tag{20.25}$$

The expression for F for this case is, as in the previous example,

$$F = -kT \ln Q(N, M, T). \tag{20.26}$$

THE TWO-DIMENSIONAL PRESSURE

In this case, before substituting for $Q(N, M, T)$ it is instructive to compare the derivative of the above expression for F with that previously determined for $F(N, V, T)$. In our previous treatments of one-phase, open chemical systems we could write

$$dF(N, V, T) = \left(\frac{\partial F}{\partial T}\right)_{N,V} dT + \left(\frac{\partial F}{\partial V}\right)_{N,T} dV + \left(\frac{\partial F}{\partial N}\right)_{V,T} dN, \tag{20.27}$$

in which, for this case, we could determine from the first and second laws that

$$\left(\frac{\partial F}{\partial T}\right)_{N,V} = -S, \quad \left(\frac{\partial F}{\partial V}\right)_{N,T} = -P, \quad \left(\frac{\partial F}{\partial N}\right)_{V,T} = \mu. \quad (20.28)$$

In the present case we may write, by analogy,

$$dF(N, M, T) = \left(\frac{\partial F}{\partial T}\right)_{M,N} dT + \left(\frac{\partial F}{\partial M}\right)_{N,T} dM + \left(\frac{\partial F}{\partial N}\right)_{M,T} dN, \quad (20.29)$$

where, as before

$$\left(\frac{\partial F}{\partial T}\right)_{M,N} = -S, \quad \left(\frac{\partial F}{\partial N}\right)_{M,T} = \mu, \quad (20.30)$$

and we define in addition

$$-\Phi \equiv \left(\frac{\partial F}{\partial M}\right)_{N,T}, \quad (20.31)$$

where Φ is effectively a two-dimensional pressure and can be expressed in terms of the partition function for the system as will be shown below.

EVALUATION OF THE THERMODYNAMIC FUNCTIONS

Given the relationships between Φ, μ, S, and F we can evaluate these parameters in terms of Q. The two-dimensional pressure, Φ, is

$$\Phi = -\left(\frac{\partial F}{\partial M}\right)_{N,T} = kT\left(\frac{\partial \ln Q}{\partial T}\right)_{N,T}$$

$$= kT \frac{\partial}{\partial M}\left[M \ln M - (M-N)\ln(M-N)\right]$$

$$= kT \ln \frac{M}{M-N} = -kT \ln\left(1 - \frac{N}{M}\right), \quad (20.32)$$

or, if we define the relative coverage, θ, as

$$\theta \equiv \frac{N}{M}, \quad (20.33)$$

we have

$$\Phi = -kT \ln(1-\theta). \quad (20.34)$$

Before we go on to evaluate S and μ, let us look at the limits on the behavior of Φ at large and small values of θ. As $\theta \to 0$, $-\ln(1-\theta) \to \theta$. Consequently,

$$\phi \to \theta kT = \left(\frac{N}{M}\right) kT. \tag{20.35}$$

This may be compared to the ideal (three-dimensional) gas equation of state

$$P = \left(\frac{N}{V}\right) kT. \tag{20.36}$$

In the other extreme, Φ increases as θ increases; and as $\theta \to 1$, its upper limit, Φ approaches ∞.

We may evaluate μ as

$$\mu = \left(\frac{\partial F}{\partial N}\right)_{M,T} = -kT \left(\frac{\partial \ln Q}{\partial N}\right)_{M,T}$$

$$= -kT \frac{\partial}{\partial N} [-N \ln N - (M-N) \ln(M-N) + N \ln q]$$

$$= kT \ln \left(\frac{\theta}{(1-\theta)q}\right). \tag{20.37}$$

Finally, we may evaluate S as

$$S = -\left(\frac{\partial F}{\partial T}\right)_{M,N} = k \ln Q + kT \left(\frac{\partial \ln Q}{\partial T}\right)_{M,N}$$

$$= k \ln \left[\frac{M!}{N!(M-N)!}\right] + Nk \ln q + NkT \left(\frac{\partial \ln q}{\partial T}\right)_{M,N}, \tag{20.38}$$

or

$$S = S_{config} + S_{vib}, \tag{20.39}$$

as in the previous case of the Einstein crystal with vacancies.

THE LANGMUIR ADSORPTION ISOTHERM

We will carry out one final manipulation with the above results, to determine the conditions for equilibrium between the adsorbed phase and the gas phase of the adsorbing species. This calculation is typical of any case in which material in one phase is in equilibrium with the same species in some other phase. The equilibrium is controlled by the criterion for distributive equilibrium deduced in our initial consideration of criteria for equilibrium in Chapter 5, namely

$$\mu^{ads} = \mu^{gas}, \tag{20.40}$$

which for this case gives

$$kT \ln\left[\frac{\theta}{(1-\theta)q_{ads}}\right] = \mu^0(T) + kT \ln\left(\frac{P}{P^0}\right). \tag{20.41}$$

(Recall from our previous statistical treatments of ideal gases that the chemical potential of such a system could always be represented by an equation of the form of the right-hand side of equation (20.41), and that, by convention, we choose P^0 to be the equilibrium partial pressure of the gas over the equilibrium condensed phase at the temperature of interest.) The above equation may be rearranged to yield

$$\ln\left[\frac{\theta}{(1-\theta)}\right] = \ln q_{ads} + \left(\frac{\mu^0(T)}{kT}\right) + \ln\left(\frac{P}{P^0}\right), \tag{20.42}$$

or

$$\frac{\theta}{(1-\theta)} = \left(\frac{P}{P^0}\right) q_{ads} e^{\mu^0(T)/kT}. \tag{20.43}$$

Finally, if we define

$$\chi(T) \equiv \left(\frac{1}{P^0}\right) q_{ads} e^{\mu^0(T)/kT}, \tag{20.44}$$

we have

$$\frac{\theta}{(1-\theta)} = \chi(T)P, \tag{20.45}$$

or

$$\theta(P, T) = \frac{\chi(T)P}{1 + \chi(T)P}. \tag{20.46}$$

This last equation is known as the *Langmuir adsorption isotherm* and gives the concentration in the adsorbed phase, in terms of θ, as a function of the

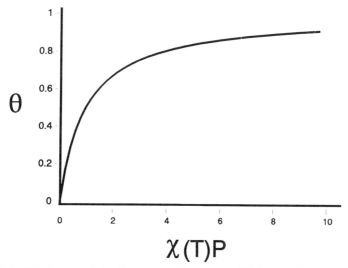

Figure 20.4 The Langmuir isotherm, showing the equilibrium adlayer coverage, θ, as a function of the gas-phase pressure, P, in terms of the temperature-dependent parameter, $\chi(T)$.

pressure in the gas phase, P, and the other parameters involved in the expression for the chemical potentials of the two phases. The form of this equation is shown in Figure 20.4. Note that at low pressures this expression reduces to

$$\theta = \chi(T)P, \qquad (20.47)$$

and the coverage initially will rise linearly with pressure. At high pressures, $\chi(T)P$ becomes large compared to unity and $\theta \to 1$ as $P \to \infty$.

BIBLIOGRAPHY

R. T. DeHoff, *Thermodynamics in Materials Science*, McGraw-Hill, New York, 1993, Chapter 13.

L. A. Girifalco, *Statistical Physics of Materials*, John Wiley & Sons, New York, 1973, Chapter 6.

T. L. Hill, *Introduction to Statistical Thermodynamics*, Addison-Wesley, 1960, Chapter 7.

E. L. Knuth, *Introduction to Statistical Thermodynamics*, McGraw-Hill, New York, 1966, Chapter 12.

J. W. Whalen, *Molecular Thermodynamics*, John Wiley & Sons, New York, 1991, Chapter 9.

PROBLEMS

20.1 Calculate the equilibrium vacancy concentration in silver at 1100 K, assuming that silver can be treated as an ideal lattice gas.

20.2 Calculate the entropy change associated with the formation of 10^{-5} mol fraction vacancies in a crystal that follows the Einstein model with $\Theta_E = 400$ K, and compare the result to the total entropy per mole at 1000 K.

20.3 Consider a system in which argon is adsorbed on carbon black to a concentration of $5 \times 10^{14}/\text{cm}^2$. The number of adsorption sites is $2 \times 10^{15}/\text{cm}^2$. If the energy of adsorption is $-12{,}000$ J/mol, the vibrational frequency is $10^{12}/\text{sec}$ for all modes, and the temperature is 100 K, what is the equilibrium pressure in the gas phase over the adsorbed layer?

20.4 Calculate the equilibrium adlayer coverage for argon adsorbed on a (111) face of copper at an argon gas pressure of 10^{-6} torr (mm of Hg). Assume that adsorption is immobile, that the energy change on adsorption is -12000 J/mol, that the argon atoms adsorb at sites of threefold coordination on the copper surface, and that the surface vibrational frequencies are $10^{13}/\text{sec}$ in the plane of the surface and $5 \times 10^{12}/\text{sec}$ normal to the surface.

20.5 A surface which has 10^{15} adsorption sites per square centimeter is exposed to a mixture of argon and krypton gases, in which $P_{Ar} = 10$ torr (mm of Hg) and $P_{Kr} = 1$ torr, at a temperature of 80 K. The adsorption energy of argon on this surface is -12000 J/mol, and that of krypton is -14000 J/mol. You may assume that the two species adsorb independently of one another and that vibrational excitation of the adsorbed species may be neglected. What are the equilibrium coverages of argon and krypton on the surface? *Hint:* The number of ways of putting N_A atoms of species A and N_B atoms of species B on M adsorption sites, with no more than one atom per site, is

$$\frac{M!}{N_A! N_B! (M - N_A - N_B)!}.$$

CHAPTER 21

THE LATTICE GAS WITH INTERACTIONS

We turn now to the case of systems in which our previous assumption that the molecules behave independently of one another is no longer valid. We will approach this subject by first considering another lattice gas system, in order to develop the methods required to handle systems of interacting particles, and then we will turn to the more complicated, but more practically important case of multicomponent condensed phase systems.

In this chapter we will consider the case of the binding of atoms to a lattice in which we take explicit account of the fact that the atoms on adjacent sites may have an energetic interaction with one another. This case is of interest in itself, in connection with the adsorption of a gas on a surface, but more importantly leads into the subject of the behavior of two component solutions, which we will discuss in Chapter 22.

THE MODEL

As in the previous case of the ideal lattice gas, we will consider a model in which molecules are bound to sites, in one, two, or three dimensions, but with the added complication that molecules on adjacent sites interact with one another. Calculation of the partition function for systems of this sort is complicated by the fact that different arrangements of the N atoms on the M sites will result in different numbers of occupied site pairs, and consequently different total energies.

250 THE LATTICE GAS WITH INTERACTIONS

The model we will use here assumes that we have a fixed number, M, of adsorption sites, with $M \to \infty$, on which are adsorbed N molecules, at a temperature T. We will develop an expression for $Q(N, M, T)$, just as in the case of the ideal lattice gas, but with the complication that while for a single molecule we can still write a contribution to the partition function $q = q(T)$, for adjacent molecules there is an interaction energy, w, which can be either attractive (negative) or repulsive (positive). The total system energy will thus depend not only on the number of molecules adsorbed and the temperature, but also on the arrangement of these molecules on the M available sites.

We can characterize the energy associated with the interacitons in terms of the number of occupied pairs of nearest-neighbor sites, which we will call N_{11}. That is, the total interaction energy is

$$E_{int} = N_{11} w. \tag{21.1}$$

We can similarly define, for future reference, N_{01} and N_{00} as the number of nearest-neighbor site pairs with only one member occupied and the number of nearest-neighbor site pairs with both sites empty. We must now develop expressions for the values of N_{11}, N_{01}, and N_{00} in a system described by the model. Let us look at an example, such as the one shown in Figure 21.1. Here we have a square lattice, with $M = 20$ and $N = 9$. The coordination number, c, of sites in this lattice (that is, the number of nearest-neighbor sites for each site) is 4. There are obviously many ways of arranging the N molecules on the M sites, two of which are shown in Figure 21.1. Many of these ways will lead to different values of N_{11}, N_{01}, and N_{00}. For a given arrangement, we may define these numbers using the following process: We draw a line from each

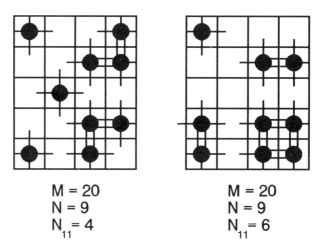

$M = 20$
$N = 9$
$N_{11} = 4$

$M = 20$
$N = 9$
$N_{11} = 6$

Figure 21.1 Schematic representation of the process used to enumerate the various types of site pairs in a system of occupied and unoccupied sites. Site pairs are classified according to the number of lines connecting the two sites of the pair.

occupied site to the four neighboring sites (as shown in the figure). For N atoms, this results in cN lines. Each 11 pair of sites will have two lines between them, each 01 pair one line, and each 00 pair no line. Thus

$$cN = 2N_{11} + N_{01}. \tag{21.2}$$

We can go through a similar line of reasoning for the $M - N$ unoccupied sites, where in this case the lines are drawn from unoccupied sites to nearest neighbor sites. This procedure leads to the equation

$$c(M - N) = 2N_{00} + N_{01}. \tag{21.3}$$

We thus have a system of fixed M and N in which we have two equations relating the three unknowns, N_{11}, N_{01}, and N_{00}. Consequently only one of these terms is independent. We shall see that it will be mathematically more convenient in the later stages of this development to make N_{01} the independent variable. In this case, we can write the total contribution to the system energy arising from interactions between adsorbed molecules as

$$N_{11}w = \left(\frac{cN}{2} - \frac{N_{01}}{2}\right)w. \tag{21.4}$$

EVALUATION OF THE PARTITION FUNCTION

Consider now how this complication of an energy per microstate that depends on the system configuration complicates the calculation of Q. In formulating the expression for Q we must, as always, account for all possible microstates of the system. In the previous case of the ideal lattice gas, this meant solving the statistical problem of the number of ways of arranging the N molecules on the M sites. In the present case, this process is complicated by the fact that different arrangements will have different numbers of 11 (or 01) site pairs, and consequently different energies. As a result, rather than looking just at the total number of possible arrangements, which we have already shown to be

$$\frac{M!}{N!(M-N)!}, \tag{21.5}$$

we must first develop an expression for the number of ways of arranging N molecules on M sites so that there are exactly, say, N_{01} site pairs of type 01, which we will call $g(N, M, N_{01})$. We must then write the expression for Q as the sum over all possible values of N_{01} of the term $g(N, M, N_{01})$, with each term multiplied by an exponential in the interaction energy associated with that value of N_{01}. That is, each term in the sum that makes up Q will contain a factor of the form

$$g(N, M, N_{01})e^{-N_{11}w/kT} = g(N, M, N_{01})\exp\left[-\frac{\left(\frac{cN}{2} - \frac{N_{01}}{2}\right)w}{kT}\right]. \quad (21.6)$$

In the expression for Q we must also account for the vibrational motion of the molecules about their equilibrium positions on the sites and the binding energy of the molecule to the site. This can be done, as in all previous cases of systems of distinguishable particles, by putting a term q^N in the expression for Q, where $q = q_0 \cdot q_v^3$ as before. The total expression for Q is thus

$$Q(N, M, T) = q^N \sum_{N_{01}} g(N, M, N_{01})\exp\left[-\frac{\left(\frac{cN}{2} - \frac{N_{01}}{2}\right)w}{kT}\right], \quad (21.7)$$

or

$$Q(N, M, T) = [qe^{-cw/2kT}]^N \sum_{N_{01}} g(N, M, N_{01})e^{-N_{01}w/kT}. \quad (21.8)$$

The next problem is to find an explicit expression for $g(N, M, N_{01})$ and then to carry out the appropriate summation. It is at this point that we must begin to make approximations in order to avoid undue mathematical complexity. The simplest approximation is to let $w \to 0$. In this limit, the exponential term in the sum for Q reduces to unity and the sum reduces to

$$\sum_{N_{01}} g(N, M, N_{01}) = \frac{M!}{N!(M-N)!}, \quad (21.9)$$

and we return to the result obtained for the lattice gas *without* interactions, as we should. This, however, is of no help in solving the problem of systems with finite interaction energy.

THE BRAGG-WILLIAMS APPROXIMATION

The first model that we will consider is the so-called *Bragg–Williams approximation*. The basic simplifying assumption made in this case is that the N atoms are arranged *randomly* on the M sites—in effect, the same assumption that is made in the definition of a *regular solution*. This is obviously an incorrect assumption, especially in systems where w/kT is large. However, it is not too bad an approximation in many real systems, especially at high temperature, and has the virtue of mathematical simplicity.

Since the atoms are arranged randomly on the sites, the appropriate factor to account for the number of possible arrangements is that developed earlier for the case of the lattice gas without interactions. The only term that must be included in Q in this case to account for the interactions is an exponential in the total interaction energy in the system, which can be written

$$e^{-\bar{N}_{11}w/kT}, \tag{21.10}$$

where $\bar{N}_{11}$ is the statistical average number of 11 site pairs in the random array of atoms on sites. We may thus write Q as

$$Q(N, M, T) = \frac{M!}{N!(M-N)!} q^N e^{-\bar{N}_{11}w/kT}. \tag{21.11}$$

The only remaining problem we are faced with is the evaluation of $\bar{N}_{11}$. This evaluation is straightforward, because the average number of near-neighbor pairs is simply the product of the number of atoms in the system, N, the number of nearest-neghbor sites that each atom has, c (the coordination number in one, two, or three dimensions), and the probability that a given near-neighbor site is occupied, which is simply N/M. This result must be divided by two to avoid counting each pair twice, yielding

$$\bar{N}_{11} = \frac{cN^2}{2M}. \tag{21.12}$$

This gives us

$$Q(N, M, T) = \frac{M!}{N!(N-M)!} q^N e^{-cN^2w/2MkT}. \tag{21.13}$$

From here we may proceed straightforwardly to the thermodynamic functions, using Stirling's approximation on the factorials. As usual we have

$$F = -kT \ln Q$$
$$= -kT\left[M \ln M - N \ln N + (M-N)\ln(M-N) + N \ln q - \frac{cN^2w}{2MkT}\right].$$
$$\tag{21.14}$$

The entropy is

$$S = k \ln Q + kT\left(\frac{\partial \ln Q}{\partial T}\right)_{N,M}$$
$$= k \ln\left[\frac{M!}{N!(M-N)!}\right] + Nk\left[\ln q + T\left(\frac{\partial \ln q}{\partial T}\right)_{M,N}\right]. \tag{21.15}$$

THE LATTICE GAS WITH INTERACTIONS

Note that this is exactly the same value deduced for the lattice gas without interactions, and shows the same subdivision into configurational and vibrational terms. This is a consequence of the assumption of random site occupancy. Continuing, we have

$$\Phi = kT\left(\frac{\partial \ln Q}{\partial M}\right)_{N,T}$$
$$= cw\theta^2 - kT\ln(1-\theta), \tag{21.16}$$

which may be compared to the result obtained for the ideal lattice gas,

$$\Phi = -kT\ln(1-\theta). \tag{21.17}$$

The only difference is the interaction energy term. Note that when w is negative (attractive interaction) the value of Φ is reduced. Finally, we have

$$\mu = -kT\left(\frac{\partial \ln Q}{\partial N}\right)_{M,T}$$
$$= kT\ln\left[\frac{\theta e^{(cw\theta/kT)}}{(1-\theta)q}\right]$$
$$= kT\ln\left(\frac{\theta}{(1-\theta)q}\right) + cw\theta. \tag{21.18}$$

This may be compared to the relation deduced for the ideal lattice gas,

$$\mu = kT\ln\left(\frac{\theta}{(1-\theta)q}\right). \tag{21.19}$$

Again, the only difference is the term involving the interaction energy.

It is also of interest to compare the isotherm equation for this case to the Langmuir isotherm developed in Chapter 20. We may proceed, as before, to equate the chemical potentials in the gas and adsorbed phases, obtaining in this case

$$kT\ln\left(\frac{\theta e^{cw\theta/kT}}{(1-\theta)q}\right) = \mu_0(T) + kT\ln\left(\frac{P}{P^0}\right). \tag{21.20}$$

This equation may be rearranged to yield

$$\left(\frac{\theta}{1-\theta}\right)e^{cw\theta/kT} = \chi(T)P, \tag{21.21}$$

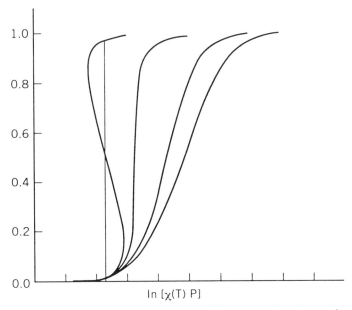

Figure 21.2 Isotherms for a lattice gas system with an attractive interaction between atoms on adjacent sites, according to the Bragg–Williams model. The system shows separation into two phases for all cases in which the parameter cw/kT is more negative than -4. This behavior may be contrasted with that observed for the Langmuir isotherm in Figure 20.4.

where

$$\chi(T) \equiv \left(\frac{1}{P^0}\right) q e^{\mu^\circ(T)/kT} \tag{21.22}$$

as before. This last result is shown graphically, for comparison with the Langmuir treatment, in Figure 21.2. Here we see that for an attractive interaction, the curve has an inflection point at $\theta = 0.5$, and becomes double valued at this coverage when cw/kT is more negative than -4, implying a first-order phase change. In such a case where the theoretical curve is double-valued, the actual observed behavior would be a vertical rise at the pressure corresponding to $\theta = 0.5$. This behavior has been observed in practice in the adsorption of gases on surfaces at low temperatures. At low gas-phase pressures a low-density adsorbed gas phase is present on the surface. As the gas-phase pressure is increased, there is an abrupt discontinuity in the isotherm at the pressure corresponding to $\theta = 0.5$, and a high-density adsorbed liquid or solid phase is formed. Measurement of the critical temperature for this condensation process permits calculation of the interaction energy.

THE QUASICHEMICAL MODEL

Now let us look at a somewhat more sophisticated model of the two- or three-dimensional lattice gas with interactions. This is the so-called quasichemical model. This model does not assume a random distribution of atoms on sites, but *does* assume that each site pair is independent of all others. This is obviously not true, especially at large values of θ. For example, for four sites of a square lattice, as shown in Figure 21.3, we have four site pairs, as shown by the four ellipses. However, the two sites of each site pair also belong to one other site pair each. The error associated with this assumption is small at low coverage and even at high coverage provides a better approximation that the Bragg–Williams model for large values of w/kT.

In treating this model we may begin by restating the general relation for Q for the lattice gas with interactions,

$$Q = q^N \sum_{N_{01}} g(N, M, N_{01}) \exp\left[-\frac{\left(\frac{cN}{2} - \frac{N_{01}}{2}\right)w}{kT}\right]. \qquad (21.23)$$

In this case, however, we may not take the configurational term outside the sum, as we did in the Bragg–Williams model, because in this case the energy of each microstate depends explicitly on the value of N_{01} for that state. We will not go into the process of evaluating the sum in equation (20.23) in detail,

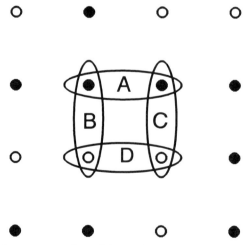

Figure 21.3 Schematic representation of the problem that arises when the assumption is made that site pairs in a lattice gas system are occupied independently of one another. In the two-dimensional square lattice shown, if the site pair A is assumed to be fully occupied, site pairs B and C cannot be fully empty.

because it is quite involved mathematically. The development is given in detail in the book by Hill referenced at the end of this chapter.

The evaluation of Q in this case leads to the relation for Φ given by

$$\Phi = kT \ln\left[\frac{(\beta + 1)(1 - \theta)}{\beta + 1 - 2\theta}\right]^{c/2} \left(\frac{1}{1 - \theta}\right), \qquad (21.24)$$

in which β is

$$\beta = [1 - 4\theta(1 - \theta)(1 - e^{-w/kT})]^{1/2}, \qquad (21.25)$$

and embodies the effect of the interaction energy.

We may finally write

$$\mu = kT\left[\ln(qe^{-cw/2kT}) + (c - 1)\ln\left(\frac{\theta}{1 - \theta}\right) + \frac{c}{2}\ln\left(\frac{\theta - \alpha}{1 - \theta - \alpha}\right)\right], \qquad (21.26)$$

where

$$\alpha = \frac{2\theta(\theta - 1)}{\beta + 1}. \qquad (21.27)$$

In this case we may again develop an expression for the adsorption isotherm, just as we have done for the ideal lattice gas and Bragg–Williams models. In this case, the resulting expression is

$$\left(\frac{\theta}{1 - \theta}\right)^{c-1} \left(\frac{1 - \theta - \alpha}{\theta - \alpha}\right)^{c/2} e^{cw/kT} = \chi(T)P, \qquad (21.28)$$

where again

$$\chi(T) = \left(\frac{1}{P^0}\right) q e^{\mu^\circ(T)/kT}. \qquad (21.29)$$

This dependence of θ on P is considerably more complicated than that deduced using the Bragg–Williams approximation. The isotherm equations for the two cases are compared in Figure 21.4. Both cases show phase changes, but the critical value of cw/kT required to produce the phase change would be larger for the quasichemical case for all values of c. The two values would approach each other as c becomes large.

We will not proceed any farther with these lattice gas models. The main point to be made here is that the models and approaches developed in this chapter will carry over into our discussion of two-component systems in Chapter 22, with the roles of filled and empty sites being taken by sites filled with two different atomic or molecular species.

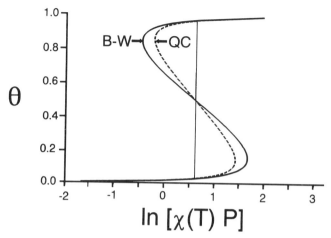

Figure 21.4 Isotherms comparing the behavior of a lattice gas system with attractive interactions between atoms on adjacent sites for the Bragg–Williams and quasichemical models. The critical value of the parameter cw/kT required to produce phase separation will be greater (more negative) for the quasichemical case.

BIBLIOGRAPHY

T. L. Hill, *Introduction to Statistical Thermodynamics*, Addison-Wesley, 1960, Chapter 14.

PROBLEMS

21.1 Suppose that in the example of the adsorption of argon on carbon black discussed in Problem 20.3, there had been an attractive lateral interaction energy between argon atoms adsorbed on adjacent sites of -1200 J/mol. If we assume that the adsorption sites form a hexagonal array, and that site occupancy is random, how much would this interaction energy change the equilibrium pressure?

21.2 The adsorption of krypton on graphite proceeds with an energy of adsorption of $-20,000$ J/mol and a lateral interaction energy of -4000 J/mol. The surface vibrational frequency is 10^{12}/sec. What gas-phase krypton pressure would be required in order to observe a first-order phase change at a temperature of 100 K? (Assume that the system can be described by the Bragg–Williams approximation and that the adsorption sites form a hexagonal array.)

21.3 When hydrogen gas at pressure P is absorbed into solid palladium, hydrogen atoms occupy interstitial sites in the FCC palladium lattice. At equilibrium we have

$$\mu_{H_2}(gas) = 2\mu_H(abs).$$

Develop an expression for the equilibrium pressure over the hydrogen/palladium solid solution, assuming that there is an interaction energy w between hydrogen atoms in adjacent interstitial sites and that the system follows the Bragg–Williams model.

21.4 Develop an expression for the energy, E, of a Bragg–Williams lattice gas, considering w to be a function of temperature, $w(T)$.

CHAPTER 22

STATISTICAL THERMODYNAMIC TREATMENT OF SOLUTIONS

We are now, finally, prepared to develop expressions for the thermodynamic behavior of multicomponent systems. We will consider explicitly two-component solutions over a wide range of composition. We will see that the results of our treatment will enable us to plot the thermodynamic properties of these systems, such as S, F, and E versus composition, and thus to demonstrate the effects that differences in intermolecular attractive forces have on these functions and on the behavior of the systems involved.

THE MODEL

We will carry out this treatment using a lattice model. This is an obvious choice for the case of a crystalline solid solution, which we will consider first, but turns out to be a reasonable approximation for simple liquid solutions as well, as we will see later in this chapter. The treatment will be similar to the one just completed for the lattice gas with interactions, and, in fact, many results of the lattice gas treatment can be taken over to the solution case with only a change in notation.

We will assume a model that consists of N_A atoms of species A plus N_B atoms of species B distributed on a lattice of $N_A + N_B$ total sites. That is, $N_A + N_B = M$ of the lattice gas treatment, and no vacant lattice sites are allowed. (We could, of course, allow for vacant sites, as we have previously, but at this point it is an unnecessary complication.) We will assume a fixed volume per site for each kind of atom in a given phase, independent of temperature.

As a consequence, for this model the volume change on mixing and the thermal expansion coefficient will both be zero, and $dF = dG$. Each atom will be assumed to vibrate about its lattice site in three dimensions. This vibrational motion can be accounted for by a three-mode vibrational partition function, q_A, for A atoms and q_B for B atoms. We will consider interactions arising from nearest neighbors only, and we will assume that the strength of the interaction depends on the kind of nearest neighbor only. That is, the well depths for the pair potentials will be ε_{AA}, ε_{BB}, and ε_{AB} for the three possible nearest-neighbor pair types. As in our previous treatments of one-component systems, these energies will be considered to be constant, positive, and independent of temperature.

We will discuss the properties of the solution in terms of the Helmholtz free energy, F, rather than G, which was used in our classical treatment of solutions. However, as explained above, the two are equivalent. Subject to the assumptions made we may write

$$dF = -SdT + \mu_A dN_a + \mu_b dN_b. \tag{22.1}$$

Note that due to the constraints placed on the system volume by the model, there is no term in dV. Also

$$F \approx G = \mu_A N_A + \mu_B N_B. \tag{22.2}$$

We may rewrite the first equation as

$$dF = -SdT + \mu_B d(N_A + N_B) + (\mu_A - \mu_B) dN_A. \tag{22.3}$$

This expression has the same form as the expression we have used in treating lattice gas systems, namely

$$dF = -SdT - \Phi dM + \mu dN, \tag{22.4}$$

where we make the association that

A atoms in the solution $\leftrightarrow$ occupied sites in the lattice gas.

We can write the expression for Q for this system as

$$Q = q_A^{N_A} q_B^{N_B} \sum_{N_{AB}} g(N_A, N_a + N_B, N_{AB}) e^{-W/kT}, \tag{22.5}$$

where

$$W = -(N_{AA}\varepsilon_{AA} + N_{AB}\varepsilon_{AB} + N_{BB}\varepsilon_{BB}), \tag{22.6}$$

and represents the total bond energy of the solution, with N_{ij} being the total number of ij bonds and ε_{ij} the pair potential for that type of bond.

This relation is essentially the same as that used in the previous lattice gas models, with

$$N_{AA} \leftrightarrow N_{11}, \quad N_{AB} \leftrightarrow N_{01}, \quad N_{BB} \leftrightarrow N_{00}, \quad N_A \leftrightarrow N, \quad N_A + N_B \leftrightarrow M. \tag{22.7}$$

Note that the function $g(N_A, N_A + N_B, N_{AB})$ is of exactly the same form as $g(N, M, N_{01})$ in the lattice gas treatment and that we can make the further analogy

$$2\varepsilon_{AB} - \varepsilon_{AA} - \varepsilon_{BB} \leftrightarrow w, \tag{22.8}$$

$$cN_A = 2N_{AA} + N_{AB} \leftrightarrow N_{11} = \frac{cN}{2} - \frac{N_{01}}{2}, \tag{22.9}$$

$$cN_B = 2N_{BB} + N_{AB} \leftrightarrow N_{00} = \frac{c(M-N)}{2} - \frac{N_{01}}{2}. \tag{22.10}$$

SOLID SOLUTIONS

For the case of a solid solution, making the substitutions shown above leads to

$$Q^S = (q_A e^{c\varepsilon_{AA}/2kT})^{N_A}(q_B e^{c\varepsilon_{BB}/2kT})^{N_B}$$
$$\cdot \sum_{N_{AB}} g(N_A, N_A + N_B, N_{AB})(e^{-w/2kT})^{N_{AB}}, \tag{22.11}$$

where the sum is exactly the same sum as in the lattice gas case. Note in passing that if we look at this in the limit $N_B \to 0$, $N_A + N_B$ = constant (in other words, as we approach pure A), the equation for Q reduces to

$$Q^S = [q_A e^{c\varepsilon_{AA}/2kT}]^{N_A}, \tag{22.12}$$

which is the same expression for Q deduced for the Einstein crystal, with

$$-c\varepsilon_{AA} \leftrightarrow u_0, \quad q_A \leftrightarrow q_v^3. \tag{22.13}$$

From this last result we can extract an expression for the chemical potential of pure solid A as

$$(\mu_A^0)^S = -kT\left(\frac{\partial \ln Q^S}{\partial N_A}\right)_T = -kT \ln(q_A e^{c\varepsilon_{AA}/2kT}). \tag{22.14}$$

One could develop a similar expression for pure B.

264 STATISTICAL THERMODYNAMIC TREATMENT OF SOLUTIONS

THE IDEAL SOLID SOLUTION

Let us now look at the evaluation of Q of the solid solution for the various types of solution that we considered in our classical treatment. Consider first the case of the ideal solid solution. In this case, since there is no net interaction between A atoms and B atoms, we must have $w = 0$ and consequently, in the expression for Q,

$$e^{-w/2kT} = 1. \qquad (22.15)$$

The sum term is thus simply

$$\sum_{N_{AB}} g(N_A, N_A + N_B, N_{AB}) = \frac{(N_A + N_B)!}{N_A! N_B!}. \qquad (22.16)$$

This leads to

$$Q^S = (q_A e^{c\varepsilon_{AA}/2kT})^{N_A} [q_B e^{c\varepsilon_{BB}/2kT}]^{N_B} \left[\frac{(N_A + N_B)!}{N_A! N_B!} \right]. \qquad (22.17)$$

From this the equations for the thermodynamic functions follow straightforwardly as

$$F^S = -kT \ln Q^S$$
$$= -N_A kT \ln(q_A e^{c\varepsilon_{AA}/2T}) - N_B kT \ln(q_B e^{c\varepsilon_{BB}/2kT})$$
$$+ kT[(N_A + N_B) \ln(N_A + N_B) - N_A \ln N_A - N_B \ln N_B]$$
$$= -\frac{N_A c\varepsilon_{AA}}{2} - \frac{N_B c\varepsilon_{BB}}{2} - N_A kT \ln q_A - N_B kT \ln q_B$$
$$+ (N_A + N_B)kT[X_A \ln X_A + X_B \ln X_B], \qquad (22.18)$$

$$E^S = kT^2 \left(\frac{\partial \ln Q^S}{\partial T} \right)_{N,V}$$
$$= kT^2 \left[-\frac{N_A c\varepsilon_{AA}}{2kT^2} + N_A \left(\frac{\partial \ln q_A}{\partial T} \right) - \frac{N_B c\varepsilon_{BB}}{2kT^2} + N_B \left(\frac{\partial \ln q_B}{\partial T} \right) \right]$$
$$= -\frac{N_A c\varepsilon_{AA}}{2} - \frac{N_B c\varepsilon_{BB}}{2} + N_A kT^2 \left(\frac{\partial \ln q_A}{\partial T} \right) + N_B kT^2 \left(\frac{\partial \ln q_B}{\partial T} \right), \qquad (22.19)$$

$$S^S = \frac{E^S}{T} - \frac{F^S}{T}$$
$$= N_A \left[kT \left(\frac{\partial \ln q_A}{\partial T} \right) + k \ln q_A \right] + N_B \left[kT \left(\frac{\partial \ln q_B}{\partial T} \right) + k \ln q_B \right]$$
$$- k(N_A + N_B)[X_A \ln X_A + X_B \ln X_B], \qquad (22.20)$$

$$\mu_A^S = -kT\left(\frac{\partial \ln Q^S}{\partial N_A}\right)_{N_B,T}$$

$$= -kT \ln[q_A e^{c\varepsilon_{AA}/2kT}] + kT \ln\left[\frac{N_A}{N_A + N_B}\right]$$

$$= -\frac{c\varepsilon_{AA}}{2} - kT \ln q_A + kT \ln X_A$$

$$= (\mu_A^0)^S + kT \ln X_A. \tag{22.21}$$

and similarly for μ_B^S.

If the solid solution is in equilibrium with a vapor phase, we will have

$$\mu_A^V = (\mu_A^0)^V + kT \ln P_A, \tag{22.22}$$

or, equating μ_A^V and μ_A^S,

$$\mu_A^S = (\mu_A^0)^V + kT \ln P_A. \tag{22.23}$$

We could also write, for the equilibrium between pure solid A and its vapor phase

$$(\mu_A^0)^S = (\mu_A^0)^V + kT \ln P_A^0. \tag{22.24}$$

If we substract the value for $(\mu_A^0)^S$, for pure solid A, from the value for μ_A^S, for A in the solid solution, we have

$$\mu_A^S - (\mu_A^0)^S = kT(-\ln q_A e^{c\varepsilon_{AA}/2kT} + \ln X_A + \ln q_A e^{c\varepsilon_{AA}/2kT})$$

$$= kT \ln X_A.$$

We may also write a similar expression in terms of the equilibrium with the vapor phase:

$$\mu_A^S - (\mu_A^0)^S = (\mu_A^0)^V + kT \ln P_A - (\mu_A^0)^V - kT \ln P_A^0. \tag{22.26}$$

Comparison of these two equations yields

$$\frac{P_A}{P_A^0} = X_A, \quad \text{or} \quad X_A = a_A, \tag{22.27}$$

as previously deduced for the ideal solution on a classical basis. Similar equations can, of course, be obtained for component B.

The changes in the thermodynamic parameters associated with the formation of the solid solution from the pure components may be evaluated by using equation (22.12) to determine

$$F_i = -kT \ln Q_i$$
$$= -\frac{N_i c \varepsilon_{ii}}{2} - N_i \ln q_i \qquad (22.28)$$

for each of the pure components, leading to

$$\Delta F_M = F^S - (F_A + F_B)$$
$$= (N_A + N_B)kT[X_A \ln X_A + X_B \ln X_B], \qquad (22.29)$$

or, on a molar basis,

$$\Delta F_M = RT[X_A \ln X_A + X_B \ln X_B]. \qquad (22.30)$$

The entropy of mixing, similarly, is

$$\Delta S_M = S^S - (S_A + s_B)$$
$$= -(N_A + N_B)k[X_A \ln X_A + X_B \ln X_B], \qquad (22.31)$$

or, on a molar basis

$$\Delta S_M = -R(X_A \ln X_A + X_B \ln X_B), \qquad (22.32)$$

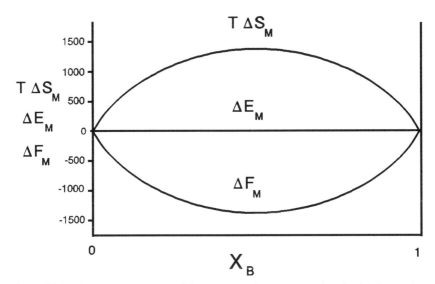

Figure 22.1 Energy, entropy, and free energy changes associated with formation of a two-component ideal solution. A single-phase solution is stable at all temperatures and compositions.

or,

$$\Delta F_M = -T\Delta S_M, \quad (22.33)$$

and consequently

$$\Delta E_M = \Delta H_M = 0 \quad (22.34)$$

again, just as deduced from classical considerations.

These relations are shown graphically in Figure 22.1, where we have plotted ΔE_M, $T\Delta S_M$, and ΔF_M versus X_B. The plot for ΔF_M shows negative values for the complete range of X_B, indicating that the solution is stable at all temperatures and compositions.

THE BRAGG–WILLIAMS MODEL AND REGULAR SOLUTIONS

We may also look at the lattice theory of solutions in terms of the Bragg–Williams model. Recall that this model assumes a finite value for the interaction energy, w, but assumes that the distribution of atoms on sites is random. As mentioned previously, this is essentially the same set of assumptions as was made in the definition of a regular solution on a classical basis. For this case, we may write, by analogy with the Bragg–Williams treatment of the lattice gas, that

$$Q^S = [q_A e^{c\varepsilon_{AA}/2kT}]^{N_A}[q_B e^{c\varepsilon_{BB}/2kT}]^{N_B}\left(\frac{(N_A+N_B)!}{N_A!N_B!}\right)e^{-\bar{N}_{AB}w/2kT}. \quad (22.35)$$

We can evaluate $\bar{N}_{AB}$, the average number of AB site pairs, by reasoning similar to that used in the lattice gas case, as the product of the number of A atoms, N_A, the number of nearest-neighbor sites, c, and the probability that a given site is acupied by a B atom, $N_B/(N_A+N_B)$. In this case, division by two is not necessary, because A and B are different species. Thus we have

$$\bar{N}_{AB} = N_A c\left(\frac{N_B}{N_a+N_B}\right) = \frac{cN_A N_B}{(N_A+N_B)}, \quad (22.36)$$

and consequently

$$Q^S = [q_A e^{c\varepsilon_{AA}/2kT}]^{N_A}[q_B e^{C\varepsilon_{BB}/2kT}]^{N_B}\left(\frac{(N_A+N_B)!}{N_A!N_B!}\right)e^{-cN_A N_B w/(N_A+N_B)2kT}, \quad (22.37)$$

or

$$\ln Q^S = N_A \ln(q_A e^{c\varepsilon_{AA}/2kT}) + N_B \ln(q_B e^{c\varepsilon_{BB}/2kT})$$
$$+ (N_A+N_B)\ln(N_A+N_B) - N_A \ln N_A - N_B \ln N_B - \frac{cN_A N_B w}{(N_A+N_B)2kT}.$$
$$(22.38)$$

From this, we may write the thermodynamic functions of the solution as

$$F^S = -kT \ln Q^S$$
$$= -N_A kT \ln(q_A e^{c\varepsilon_{AA}/2kT}) - N_B kT \ln(q_B e^{c\varepsilon_{BB}/2kT})$$
$$+ kT[(N_A + N_B)\ln(N_A + N_B) - N_A \ln N_A - N_B \ln N_B] + \frac{cwN_A N_B}{2(N_A + N_B)}$$
$$= -\frac{N_A c\varepsilon_{AA}}{2} - \frac{N_B c\varepsilon_{BB}}{2} - N_A kT \ln q_A - N_B kT \ln q_B$$
$$+ (N_A + N_B)kT[X_A \ln X_A + X_B \ln X_B] + \frac{cwN_A N_B}{2(N_A + N_B)}, \qquad (22.39)$$

$$E^S = kT^2 \left(\frac{\partial \ln Q}{\partial T}\right)_{N,V}$$
$$= kT^2 \left[-\frac{N_A c\varepsilon_{AA}}{2kT^2} + N_A\left(\frac{\partial \ln q_A}{\partial T}\right) - \frac{N_B c\varepsilon_{BB}}{2kT^2} + N_B\left(\frac{\partial \ln q_B}{\partial T}\right) - \frac{cwN_A N_B}{2(N_A+N_B)kT^2}\right]$$
$$= -\frac{N_A c\varepsilon_{AA}}{2} - \frac{N_B c\varepsilon_{BB}}{2} + N_A kT^2\left(\frac{\partial \ln q_A}{\partial T}\right) + N_B kT^2\left(\frac{\partial \ln q_B}{\partial T}\right) - \frac{cwN_A N_B}{2(N_A+N_B)}$$
$$\qquad (22.40)$$

$$S^S = \frac{E^S}{T} - \frac{F^S}{T}$$
$$= N_A\left[kT\left(\frac{\partial \ln q_A}{\partial T}\right) + k \ln q_A\right] + N_B\left[kT\left(\frac{\partial \ln q_B}{\partial T}\right) + k \ln q_B\right]$$
$$- k(N_A + N_B)[X_A \ln X_A + X_B \ln X_B], \qquad (22.41)$$

$$\mu_A^S = -kT\left(\frac{\partial \ln Q}{\partial N_A}\right)_{N_B,T}$$
$$= -kT \ln\left[q_A \frac{e^{c\varepsilon_{AA}}}{2kT}\right] + kT \ln\left[\frac{N_A}{N_A + N_B}\right]$$
$$- kT \frac{\partial}{\partial N_A}\left(\frac{cwN_A N_B}{(N_A + N_B)2kT}\right)$$
$$= -\frac{c\varepsilon_{AA}}{2} - kT \ln q_A + kT \ln X_A - \frac{cw(1 - X_A)^2}{2}. \qquad (22.42)$$

and similarly for μ_B^S.

Note that the expressions for F, E, and μ differ from the corresponding expressions for the ideal solution, developed earlier, only in the final term, which is associated with the interaction energy.

We may use these expressions, along with the expressions previously developed for the pure components, to determine the changes in F, S, and E associated with mixing as

$$\Delta F_M = F^S - F_A - F_B$$

$$= -kT[(N_A + N_B)\ln(N_A + N_B) + N_A \ln N_A + N_B \ln N_B] - \frac{cw N_A N_B}{2(N_A + N_B)}$$

$$= kT\left[N_A \ln \frac{N_A}{N_A + N_B} + N_B \ln \frac{N_B}{N_A + N_B}\right] - \frac{cw N_A N_B}{2(N_A + N_B)}, \quad (22.43)$$

or, by multiplying through by $N_{Av}/(N_A + N_B)$ to put the result on a molar basis,

$$\Delta F_M = N_{Av} kT[X_A \ln X_A + X_B \ln X_B] - N_{Av} \frac{cw}{2} X_A X_B$$

$$= RT(X_A \ln X_A + X_B \ln X_B) - N_{Av} \frac{cw}{2} X_A X_B. \quad (22.44)$$

Similarly

$$\Delta S_M = S^S - S_A - S_B$$

$$= -k(N_A \ln X_A + N_B \ln X_B), \quad (22.45)$$

or, on a molar basis,

$$\Delta S_M = -R(X_A \ln X_A + X_B \ln X_B), \quad (22.46)$$

which is identical to the result derived for the ideal solution, as it should be. The molar energy change on mixing is simply

$$\Delta E_M = \Delta F_M + T\Delta S_M$$

$$= -N_{Av} \frac{cw}{2} X_A X_B. \quad (22.47)$$

Note that since we have assumed $w \neq f(T)$, the expression for E_M is temperature-independent.

Again we may look at the properties of the solution graphically. In this case, the results will look different depending on whether w is positive or negative. For the case of w positive (unlike pairs favored), the appropriate curves of ΔE_M, ΔS_M, and ΔF_M versus composition are shown in Figure 22.2. As in the case of the ideal solution, ΔF_M is always negative and the system is stable as a

270 STATISTICAL THERMODYNAMIC TREATMENT OF SOLUTIONS

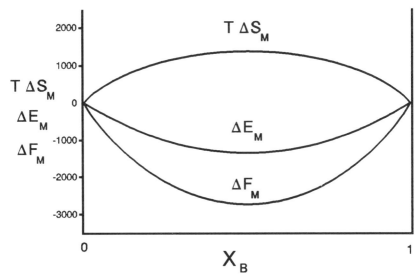

Figure 22.2 Energy, entropy, and free energy changes associated with formation of a nonideal two-component solution, in which there is a net attractive interaction between A and B atoms on adjacent sites, according to the Bragg–Williams model. A single-phase solution is stable at all temperatures and compositions.

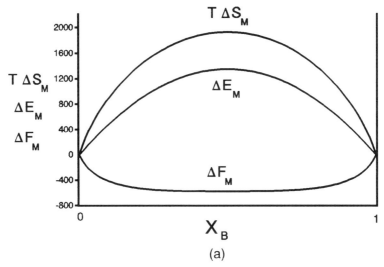

(a)

Figure 22.3 Energy, entropy, and free energy changes associated with the formation of a nonideal two-component solution in which there is a net repulsive interaction between A and B atoms on adjacent sites, according to the Bragg–Williams model. (a) At high temperatures, a single-phase solution is stable at all compositions. (b and c) At lower temperatures, phase separation occurs, leading to two phases whose compositions are defined by the common tangent to the two minima in the free energy curve.

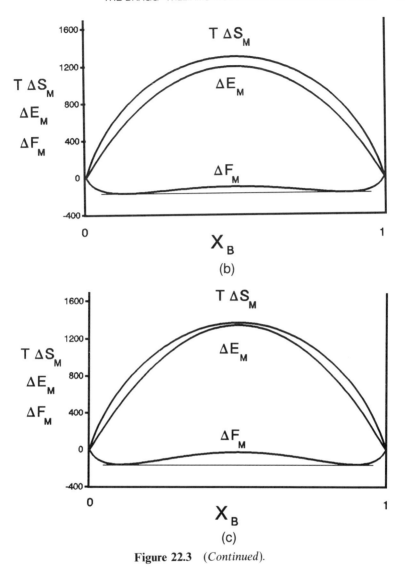

Figure 22.3 (*Continued*).

one-phase random solution over the entire composition range. In the case of w negative (like pairs favored) the curves, as shown in Figure 22.3, are quite different. In this case, at low temperatures, where the effect of the $T\Delta S_M$ term is smallest, the positive ΔE_M term causes the ΔF_M curve to rise in the middle of the composition range. This leads to phase separation. At equilibrium, the system will separate into two random solution phases. The compositions of these two phases are set by the usual criterion for distributive equilibrium, namely

$$\mu_i^\alpha = \mu_i^\beta. \tag{22.48}$$

272 STATISTICAL THERMODYNAMIC TREATMENT OF SOLUTIONS

The location of these compositions is determined by the tangent line connecting the two low points in the curve of ΔF_M versus composition. At these points $\mu_i^\alpha = \mu_i^\beta$, as one can determine using the treatment developed in Chapter 18 relating the partial molar properties to the total molar properties.

THE QUASICHEMICAL MODEL

As a final means of characterizing solutions, we may apply the quasichemical model to the case of a two-component solution. We will not go through the full derivation of Q in this case, but will reason by direct analogy with the quasichemical treatment of the lattice gas. For the present case we will have

$$\ln Q^S = N_A \ln(q_A e^{c\varepsilon_{AA}/2kT}) + N_B \ln(q_B e^{c\varepsilon_{BB}/2kT})$$
$$+ \ln \sum_{N_{AB}} g[N_a, N_A + N_B, N_{AB}] e^{-w N_{AB}/2kT}. \quad (22.49)$$

Recalling the analogy made earlier in this selection, we see that

$$\mu_B^S d(N_A + N_B) \leftrightarrow -\Phi dM, \quad (22.50)$$

or

$$\left(\frac{\partial F}{\partial(N_A + N_B)}\right)_{N,T} \leftrightarrow \left(\frac{\partial F}{\partial M}\right)_{N,T}, \quad (22.51)$$

or

$$\mu_B^S = -kT \left[\frac{\partial}{\partial(N_A + N_B)} (N_A \ln q_A e^{c\varepsilon_{AA}/2kT} + N_B \ln q_B e^{c\varepsilon_{BB}/2kT})\right]$$
$$- kT \left[\frac{\partial}{\partial(N_A + N_B)} \left(\ln \sum_{N_{AB}} g[N_A, N_A + N_B, N_{AB}] e^{-w N_{AB}/2kT}\right)\right]. \quad (22.52)$$

We will not carry out this differentiation, simply state the result, which is

$$\mu_B^S = -kT \ln[q_B e^{c\varepsilon_{BB}/2kT}] + kT \ln X_B - kT \left(\frac{c}{2}\right) \ln \frac{(\beta + 1)X_B}{\beta - 1 + 2X_B}, \quad (22.53)$$

where β, by analogy with the lattice gas treatment, is

$$\beta = [1 - 4X_A X_B (1 - e^{-w/kT})]^{1/2}. \quad (22.54)$$

A similar expression can be obtained for μ_A^S. Note in passing that if we consider the equilibrium between the solution and the corresponding vapor phase we have

$$\mu_i^S = \mu_i^V$$

$$(\mu_i^0)^S + kT \ln X_i - kT \left(\frac{c}{2}\right) \ln \frac{(\beta+1)X_i}{\beta - 1 + 2X_i} = (\mu_i^0)^V + kT \ln \frac{P_i}{P_i^0}. \quad (22.55)$$

This may be compared to the expression developed classically for a concentrated solution, namely

$$\frac{P_i}{P_i^0} = \gamma X_i, \quad (22.56)$$

to show that

$$\left(\frac{\beta - 1 + 2X_i}{(\beta+1)X_i}\right)^{c/2} = \gamma_i. \quad (22.57)$$

Finally, let us consider the terms contributing to ΔF_M for this case. In general, we may write

$$F^S = N_{Av}(X_A \mu_A + X_B \mu_B)$$
$$= RTX_A \left(-\ln q_A e^{c\varepsilon_{AA}/2kT} + \ln X_A - \left(\frac{c}{2}\right) \ln \frac{(\beta+1)X_A}{\beta - 1 + 2X_A}\right)$$
$$+ RTX_B \left(-\ln q_B e^{c\varepsilon_{BB}/2kT} + \ln X_B - \left(\frac{c}{2}\right) \ln \frac{(\beta+1)X_B}{\beta - 1 + 2X_B}\right). \quad (22.58)$$

We may also write as usual

$$\Delta F_M = F^S - X_A F_A - X_B F_B, \quad (22.59)$$

which leads, in this case, to

$$\Delta F_M = RT(X_a \ln X_A + X_B \ln X_B)$$
$$+ RT\left(\frac{c}{2}\right)\left[X_A \ln \frac{\beta - 1 + 2X_A}{(\beta+1)X_A} + X_B \ln \frac{\beta - 1 + 2X_B}{(\beta+1)X_B}\right], \quad (22.60)$$

which differs from the ideal case only in the last term.

By similar reasoning we may deduce that

$$\Delta E_M = -N_{Av}\left(\frac{cwX_A X_B}{\beta + 1}\right). \quad (22.61)$$

These two results can be expanded in a power series in w/kT to yield the approximate results

$$\Delta F_M \approx RT(X_A \ln X_A + X_B \ln X_B)$$
$$- RT\left(\frac{X_A X_B}{2}\right)\left(\frac{cw}{kT}\right) - RT\left(\frac{X_A^2 X_B^2}{4c}\right)\left(\frac{cw}{kT}\right)^2 + \cdots \quad (22.62)$$

and

$$\Delta E_M \approx - RT\left(\frac{X_A X_B}{2}\right)\left(\frac{cw}{kT}\right) - RT\left(\frac{X_A^2 X_B^2}{2c}\right)\left(\frac{cw}{kT}\right)^2 + \cdots. \quad (22.63)$$

These results lead to

$$\Delta S_M \approx - R(X_a \ln X_A + X_B \ln X_B) - R\left(\frac{X_A^2 X_B^2}{4c}\right)\left(\frac{cw}{kT}\right)^2 + \cdots. \quad (22.64)$$

This also differs from the ideal case only in the last term. Note that ΔS_M for this case will be *less* than the ideal entropy of mixing for all possible values of w, whether w is positive or negative, as w enters as a squared term. This is as

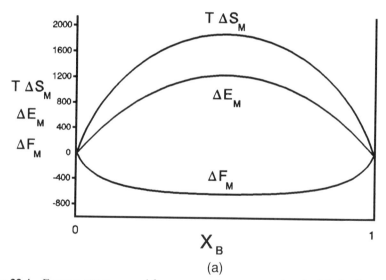

(a)

Figure 22.4 Energy, entropy, and free energy changes associated with the formation of nonideal solid solutions in which there is a net repulsive interaction between A and B atoms on adjacent sites, according to the quasichemical treatment. (a) At high temperatures, a random, single-phase solution is stable at all compositions. (b) At a lower temperature, the system reaches a critical temperature at which the second derivative of the free energy curve goes to zero. (c) At still lower temperatures, phase separation occurs, leading to two phases whose compositions are given by the common tangent to the minima in the free energy curve.

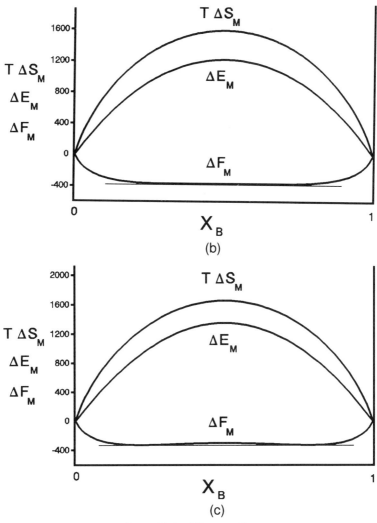

Figure 22.4 (*Continued*).

expected, because any departure from randomness, either to favor AB pairs or to favor like pairs, will reduce the number of possible arrangements of the A and B atoms and will consequently reduce the entropy.

These results may be shown graphically by plotting ΔF_M, ΔE_M, and $T\Delta S_M$ versus composition as in the previous cases. For the case of w negative (like pairs favored) the resulting curves look very similar to those obtained for the Bragg–Williams model, as shown in Figure 22.4. The principal difference is that phase separation occurs at lower temperatures. The alternate case, w positive (AB pairs favored), leads to a new class of behavior, namely ordering. This is shown in Figure 22.5. Here, at low enough temperature the energy

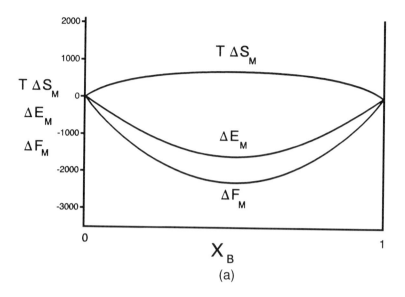

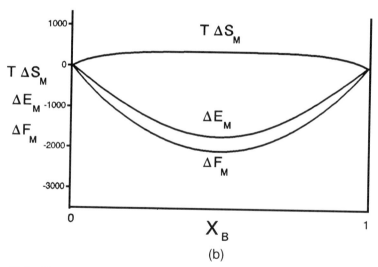

Figure 22.5 Energy, entropy, and free energy changes associated with the formation of a nonideal two-component solution in which there is a net attractive interaction between A and B atoms on adjacent sites, according to the quasichemical model. At all temperatures and compositions, a single phase solution is stable. (a) At high temperatures, the solution will be random at all compositions. (b, c) At lower temperatures, the solution will be partially (b) or completely (c) ordered, which will greatly reduce the entropy of mixing.

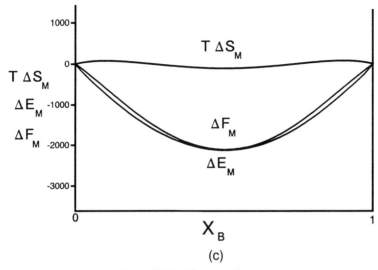

Figure 22.5 (*Continued*).

decrease associated with maximizing the number of AB pairs outweighs the unfavorable decrease in the entropy associated with increased order, and an ordered one-phase solution results.

LIQUID SOLUTIONS

As was mentioned early in this chapter, the lattice theory we have just used to develop expressions for the behavior of solid solutions can also be used for liquid solutions, provided that account is taken of the fundamental differences between solids and liquids. In order to adapt this model, we must take account of three factors: the difference in binding energy associated with the different average interatomic distance in the liquid, the difference in the vibrational energy terms, and the greatly increased atomic mobility in the liquid. We will do this by extending the "cell" model of the liquid developed in Chapter 16 to the case of a two-component liquid solution.

The difference in binding energy will be handled as before by assuming that the binding energy in the liquid solution can be represented by a "relaxed" interatomic potential, using the same potential energy curves for AA, BB, and AB bonds as was used for the solid solution, but with an interatomic distance consistent with the density of the liquid solution. We will also introduce the simplification that the density of the liquid solution varies linearly with composition, or, in other words, that the ΔV_M for the liquid solution is zero. This implies that the AA, BB, and AB bond lengths, and consequently the bond

278 STATISTICAL THERMODYNAMIC TREATMENT OF SOLUTIONS

energies, will be independent of composition. We may thus represent the bond energies as ε'_{AA}, ε'_{BB}, and ε'_{AB}, where the exact relation between the ε_{ij} for the solid and ε'_{ij} for the liquid will depend on the form chosen for the interatomic potential.

The vibrational energy terms will also be handled as in the previous case of pure liquids, where we will associate a q'_i with each atom of the form

$$q'_i = \left(\frac{kT}{hv_l}\right)^3, \tag{22.65}$$

and we may relate $(v_l)_i$ to $(v_s)_i$ through the form of the interatomic potential as was done in the case of one component systems in Chapter 17.

Finally, the entropy difference associated with the increased mobility of the liquid will be treated exactly as was done in the case of the pure liquid by including an additional factor of e in the expression for q, following the same logic that was used in equations (16.4) to (16.6).

Again, it should be noted that this set of assumptions is by no means the only possible choice. As in our previous models, it has been made in order to produce a model that reproduces the general features of the behavior of real systems without introducing inordinate mathematical complications.

THE PARTITION FUNCTION

Subject to the assumptions stated above, the partition function for the liquid solution may be written as

$$Q^L = (q'_A e^{c\varepsilon'_{AA}/2kT})^{N_A}(q'_B e^{c\varepsilon'_{BB}/2kT})^{N_B} e^{(N_A+N_B)}$$
$$\cdot \sum_{N_{AB}} g(N_A, N_A+N_B, N_{AB})(e^{-w'/2kT})^{N_{AB}}, \tag{22.66}$$

where, in this case,

$$w' = 2\varepsilon'_{AB} - \varepsilon'_{AA} - \varepsilon'_{BB}. \tag{22.67}$$

Note that for the case in which $N_B \to 0$, we have

$$Q^L = (q'_A e^{c\varepsilon'_{AA}/2kT} e)^{N_A}, \tag{22.68}$$

or

$$\ln Q^L = N_A\left(\frac{c\varepsilon'_{AA}}{2kT} + \ln(q'_A) + 1\right), \tag{22.69}$$

which, if we recall that

$$q'_A = \left(\frac{kT}{hv_l}\right)^3_A, \tag{22.70}$$

is exactly the same as the expression developed for the partition function of the pure liquid in Chapter 16. As in the case of the solid solution, a similar equation can be written for pure liquid B.

IDEAL LIQUID SOLUTIONS

The difficulties associated with evaluating the sum term in equation (22.66) are the same in this case as for the solid solution, and can be handled similarly. As before, we may consider the case in which $w' = 0$, implying no net interaction between A and B atoms. This case again leads to

$$Q^L = (q'_A e^{c\varepsilon'_{AA}/2kT})^{N_A}(q'_B e^{c\varepsilon'_{BB}/2kT})^{N_B}\frac{(N_A + N_B)!}{N_A!N_B!}e^{(N_A+N_B)} \quad (22.71)$$

and differs from the corresponding expression for the solid solution only in the term associated with the "communal entropy." The expressions for the thermodynamic functions follow straightforwardly, leading to

$$F^L = -kT \ln Q^L$$

$$= -\frac{N_A c\varepsilon'_{AA}}{2} - \frac{N_B c\varepsilon'_{BB}}{2} - N_A kT \ln q'_A - N_B kT \ln q'_B$$

$$+ (N_A + N_B)kT[X_A \ln X_A + X_B \ln X_B] - (N_A + N_B)kT, \quad (22.72)$$

$$E^L = kT^2\left(\frac{\partial \ln Q^L}{\partial T}\right)_{N,V}$$

$$= -\frac{N_A c\varepsilon'_{AA}}{2} - \frac{N_B c\varepsilon'_{BB}}{2} + N_A kT^2\left(\frac{\partial \ln q'_A}{\partial T}\right) + N_B kT^2\left(\frac{\partial \ln q'_B}{\partial T}\right), \quad (22.73)$$

$$S^L = \frac{E^L}{T} - \frac{F^L}{T}$$

$$= N_A\left[kT\left(\frac{\partial \ln q'_A}{\partial T}\right) + k \ln q'_A\right] + N_B\left[kT\left(\frac{\partial \ln q'_B}{\partial T}\right) + k \ln q'_B\right]$$

$$- k(N_A + N_B)[X_A \ln X_A + X_B \ln X_B] + (N_A + N_B)k, \quad (22.74)$$

$$\mu_A^L = -kT\left(\frac{\partial \ln Q^L}{\partial N_A}\right)_{N_B,T}$$

$$= -\frac{c\varepsilon'_{AA}}{2} - kT \ln q'_A + kT \ln X_A - kT$$

$$= (\mu_A^0)^L + kT \ln X_A, \quad (22.75)$$

and similarly for μ_B^L.

The changes in the thermodynamic functions accompanying mixing, per mole, are thus

$$\Delta F_M = RT(X_A \ln X_A + X_B \ln X_B),$$
$$\Delta E_M = 0, \qquad (22.76)$$
$$\Delta S_M = -R(X_A \ln X_A + X_B \ln X_B),$$

which are identical to those developed for the solid solution, and will lead to the same behavior in terms of ΔF_M versus X_B as was shown for the solid solution in Figure 22.1.

REGULAR LIQUID SOLUTIONS

The regular solution model is in general a very good model for liquid solutions, because the high atomic mobility mitigates against ordering or clustering on a long-term basis. Here again, for the model we have chosen, we may express the energy associated with mixing in terms of a statistical average number of AB bonds, using

$$\bar{N}_{AB} = \frac{cN_A N_B}{(N_A + N_B)}. \qquad (22.77)$$

This leads to

$$Q^L = (q'_A e^{c\varepsilon'_{AA}/2kT})^{N_A}(q'_B e^{c\varepsilon'_{BB}/2kT})^{N_B}\left(\frac{(N_A+N_B)}{N_A! N_B!}\right)$$
$$\times e^{-cN_A N_B w'/(N_A+N_B)kT} e^{N_A+N_B}, \qquad (22.78)$$

where again the form differs from the corresponding equation for the solid solution only in the "communal entropy" term.

The relations for the thermodynamic functions in this case are

$$F^L = -kT \ln Q^L$$
$$= -\frac{N_A c\varepsilon'_{AA}}{2} - \frac{N_B c\varepsilon'_{BB}}{2} - N_A kT \ln q'_A - N_B kT \ln q'_B$$
$$+ (N_A + N_B)kT[(X_A \ln X_A + X_B \ln X_B) - 1] + \frac{cwN_A N_B}{2(N_A + N_B)}, \qquad (22.79)$$

$$E^L = kT^2\left(\frac{\partial \ln Q^L}{\partial T}\right)_{N,V}$$

$$= -\frac{N_A c\varepsilon'_{AA}}{2} - \frac{N_B c\varepsilon'_{BB}}{2} + N_A kT^2\left(\frac{\partial \ln q'_A}{\partial T}\right)$$

$$+ N_B kT^2\left(\frac{\partial \ln q'_B}{\partial T}\right) - \frac{cw' N_A N_B}{2(N_A + N_B)}, \qquad (22.80)$$

$$S^L = \frac{E^L}{T} - \frac{F^L}{T}$$

$$= N_A\left[kT\left(\frac{\partial \ln q'_A}{\partial T}\right) + k\ln q'_A\right] + N_B\left[kT\left(\frac{\partial \ln q'_B}{\partial T}\right) + k\ln q'_B\right]$$

$$- k(N_A + N_B)[(X_A \ln X_A + X_B \ln X_B) - 1], \qquad (22.81)$$

$$\mu_A^L = -kT\left(\frac{\partial \ln Q^L}{\partial N_A}\right)_{N_B, T}$$

$$= -\frac{c\varepsilon'_{AA}}{2} - kT \ln q'_A + kT \ln X_A + kT - \frac{cw(1 - X_A)^2}{2}, \qquad (22.82)$$

and similarly for μ_B.

The corresponding expressions for the changes associated with mixing have exactly the same form as those deduced for the regular solid solution in equations (22.44), (22.45), and (22.46), with w' substituted for w, and will lead to ΔF_M versus X_B curves having the same form as those shown in Figure 22.2, for $w' > 0$, and Figure 22.3, for $w' < 0$.

We will not carry the case of liquid solutions any further. We could, of course, develop a model similar to the quasichemical model developed for the solid solution case. This development would not add much to our understanding of liquid solutions, due to the lack of persistent ordering or clustering in the liquid state.

BIBLIOGRAPHY

R. T. DeHoff, *Thermodynamics in Materials Science*, McGraw-Hill, New York, 1993, Chapter 8.

L. A. Girifalco, *Statistical Physics of Materials*, John Wiley & Sons, New York, 1973, Chapter 5.

T. L. Hill, *Introduction to Statistical Thermodynamics*, Addison-Wesley, 1960, Chapter 20.

R. A. Swalin, *Thermodynamics of Solids*, John Wiley & Sons, New York, 1962, Chapter 9.

J. W. Whalen, *Molecular Thermodynamics*, John Wiley & Sons, New York, 1991, Chapter 8.

PROBLEMS

22.1 Develop an expression for the critical temperature for phase separation in a two-component system that forms a regular solution. Evaluate this expression for an alloy in which the energy of sublimation of pure A is 50,000 J/mol, the energy of sublimation of pure B is 60,000 J/mol, both A and B are face-centered cubic in their pure forms, and the strength of an AB bond is 0.9 times the strength of an AA bond.

22.2 Calculate the entropy of an alloy consisting of a random solid solution of 1% aluminum in silver at 500 K. Assume that the energy of vaporization of aluminum from the alloy is the same as that of silver. State and justify any other assumptions that you make.

22.3 Calculate the difference in C_V between pure silver and a silver alloy containing 2% aluminum at 100 K and at 1000 K, using the Einstein model. Neglect the electronic contribution to C_V, and use $\Theta_E = 160$ K for silver and 300 K for aluminum.

22.4 Calculate the free energy change for the formation of an FCC solid solution of A and B, in which $\varepsilon_{AA} = 50,000$ J/mol, $\varepsilon_{BB} = 60,000$ J/mol, and $\varepsilon_{AB} = 58,000$ J/mol, at a composition of $X_B = 0.4$ and a temperature of 800 K:

(a) Assuming that the system follows the Bragg–Williams model.

(b) Assuming that the system follows the quasichemical model.

22.5 For the system in Problem 22.4, calculate the difference in free energy of formation of solid and liquid solutions at $X_B = 0.4$ and $T = 800$ K, assuming that the solid follows the Bragg–Williams model, that the liquid is ideal, that the interatomic potentials are described by the Lennard-Jones 6-12 potential, and that $\rho_l = 0.95\rho_s$.

CHAPTER 23

PHASE EQUILIBRIUM IN MULTICOMPONENT SYSTEMS

In this chapter we will use the criterion for distributive equilibrium developed in Chapter 5, the general equations describing the behavior of solutions developed in Chapter 19, and the statistical mechanical treatment of solutions developed in Chapter 22 to construct two-component phase diagrams for systems having a wide range of behaviors. We will see that the general features observed in real systems can be reproduced by suitable choices of model parameters.

THE MODEL

We will treat the two-component condensed phases involved using the expressions developed in Chapter 22 for solid and liquid binary systems. To limit the mathematical complexity involved, we will confine our treatments to systems having (a) a single, ideal liquid phase and (b) a solid phase that may show either ideal or regular solution behavior. In determining the equilibrium vapor pressure over the condensed phases, we will assume that the vapor phase is an ideal gas mixture. Note that these restrictions are made only for computational convenience. There is no reason why the techniques used here could not be used for more complicated systems, say for liquid solutions that were nonideal and thus showed phase separation in the liquid, or solid solutions that could be modelled using the quasichemical approach that would show ordering at low temperatures.

We will in all cases use the general criterion for distributive equilibrium, namely

$$\mu_i^\alpha = \mu_i^\beta = \mu_i^\gamma = \cdots . \qquad (23.1)$$

That is, for a two-component system in which we have a solid phase in equilibrium with a liquid phase we would have

$$\mu_A^S = \mu_A^L, \qquad \mu_B^S = \mu_B^L. \qquad (23.2)$$

THE TWO-COMPONENT IDEAL SYSTEM

We will begin by considering equilibria between solid and liquid phases, and we will treat equilibria involving gas phases later in this chapter. Consider first the simplest possible case, namely a system in which the only stable condensed phases are an ideal solid solution and an ideal liquid solution. For this case we may write

$$\mu_i^\alpha = (\mu_0)_i^\alpha + kT \ln X_i^\alpha \qquad (23.3)$$

for each component in each phase. We may further write expressions for the molar Helmholz free energy of each phase of the form

$$F^\alpha = X_A N_{Av} \mu_A^\alpha + X_B N_{Av} \mu_B^\alpha, \qquad (23.4)$$

or, using equation (23.3),

$$F^\alpha = X_A N_{Av} (\mu_A^0)^\alpha + X_B N_{Av} (\mu_B^0)^\alpha + RT(X_A \ln X_A + X_B \ln X_B), \qquad (23.5)$$

or, making the substitution

$$X_A = (1 - X_B), \qquad (23.6)$$

we have

$$F^\alpha = (1 - X_B) N_{Av} (\mu_A^0)^\alpha + X_B N_{Av} (\mu_B^0)^\alpha + RT[1 - X_B) \ln(1 - X_B) + X_B \ln X_B] \qquad (23.7)$$

The forms of the resulting curves for the solid and liquid phases at various temperatures are shown schematically in Figure 23.1 for a system in which the melting point of pure A, T_{m_A}, is greater than that of pure B, T_{m_B}. For $T > T_{m_A}$ the curve for the solid phase lies completely above that for the liquid phase, and the liquid is stable at all values of X_B. Conversely, at $T < T_{m_B}$ the solid curve lies below that of the liquid, and the solid phase is stable at all X_B. In the intermediate region, $(T_m)_A > T > (T_m)_B$, the curves must cross at least once, because the stable phase of pure A is solid and that of pure B is liquid. In this region we may use the criterion for distributive equilibrium to determine the equilibrium values of X_B^L and X_B^S at any temperature.

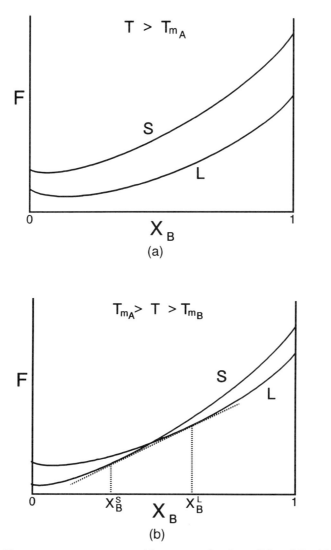

Figure 23.1 Free energy versus composition curves for the solid and liquid phases in a two-component system in which $T_{m_A} > T_{m_B}$, showing ideal behavior in both phases. (a) $T_1 > T_{m_A}$. (b) $T_{m_A} > T_2 > T_{m_B}$. (c) $T_3 < T_{m_B}$. In the range between T_{m_A} and T_{m_B}, solid and liquid phases will coexist, with compositions defined by the common tangent to the two free energy curves. (*Continued on next page.*)

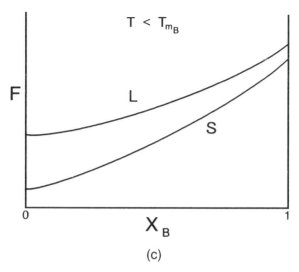

(c)

Figure 23.1 (*Continued*)

If we equate μ_B^L and μ_B^S, we have

$$\mu_B^S = \mu_B^L$$
$$(\mu_B^0)^S + kT \ln X_B^S = (\mu_B^0)^L + kT \ln X_B^L, \tag{23.8}$$

or

$$\ln \frac{X_B^S}{X_B^L} = \frac{(\mu_B^0)^L - (\mu_B^0)^S}{kT}. \tag{23.9}$$

By similar reasoning we may obtain

$$\ln \frac{X_A^S}{X_A^L} = \frac{(\mu_A^0)^L - (\mu_A^0)^S}{kT}, \tag{23.10}$$

or

$$\ln \frac{1 - X_B^S}{1 - X_B^L} = \frac{(\mu_A^0)^L - (\mu_A^0)^S}{kT}. \tag{23.11}$$

We now have two equations in two unknowns, provided that we can evaluate the $(\mu_i^0)^\alpha$ terms either experimentally or in terms of a model. Solution of equations (23.9) and (23.11) for X_B^S and X_B^L yields

$$X_B^S = \frac{1 - e^{-(\Delta\mu_A^0)/kT}}{e^{-(\Delta\mu_B^0)/kT} - e^{-(\Delta\mu_A^0)/kT}},$$
$$X_B^L = X_B^S e^{-(\Delta\mu_B^0)/kT}, \tag{23.12}$$

where

$$(\Delta\mu_i^0) = (\mu_i^0)^L - (\mu_i^0)^S$$
$$= \frac{1}{N_{Av}}[(\Delta E_f)_i - T(\Delta S_f)_i], \qquad (23.13)$$

where ΔE_f and ΔS_f are the molar energy and entropy of fusion of pure i and we are using the approximation that $E \approx H$ for condensed phases.

To evaluate these relations, we must evaluate the $(\Delta\mu_i^0)$ terms. Each of these terms is simply the difference in chemical potential between liquid and solid phases for the pure component i at any temperature. At the melting point of pure i we have

$$(\Delta\mu_i^0) = 0, \qquad (23.14)$$

and thus we have

$$(\Delta E_f)_i = T_{m_i}(\Delta S_f)_i. \qquad (23.15)$$

leading to

$$N_{Av}(\Delta\mu_i^0) = (\Delta S_f)_i(T_{m_i} - T). \qquad (23.16)$$

Substituting this relation into equations (23.12) yields

$$X_B^S = \frac{1 - \exp\left(-\dfrac{(\Delta S_f)_A(T_{m_A} - T)}{RT}\right)}{\exp\left(-\dfrac{(\Delta S_f)_B(T_{m_B} - T)}{RT}\right) - \exp\left(-\dfrac{(\Delta S_f)_A(T_{m_A} - T)}{RT}\right)}$$

$$X_B^L = X_B^S \exp\left(-\frac{(\Delta S_f)_B(T_{m_B} - T)}{kT}\right). \qquad (23.17)$$

At this point, one could proceed experimentally using measured values of the $(\Delta S_f)_i$ and T_{m_i} terms. Alternatively, we may use the statistical mechanical expressions for the μ_i^0 developed in Chapters 14 and 16, along with models for the solid and liquid. For this case

$$\Delta\mu_B^0 = (\mu_B^0)^L - (\mu_B^0)^S$$
$$= \frac{(u_{0_B})^L}{2} - 3kT\ln\left(\frac{kT}{h\nu_{l_B}}\right) - kT - \frac{(u_{0_B})^S}{2} + 3kT\ln\left(\frac{kT}{h\nu_{S_B}}\right)$$
$$= \frac{(u_{0_B})^L - (u_{0_B})^S}{2} - kT\left[3\ln\left(\frac{\nu_{S_B}}{\nu_{l_B}}\right) + 1\right], \qquad (23.18)$$

and similarly

$$\Delta\mu_A^0 = \frac{(u_{0_A})^L - (u_{0_A})^S}{2} - kT\left[3\ln\left(\frac{v_{s_A}}{v_{l_A}}\right) + 1\right]. \quad (23.19)$$

We may relate the $\Delta\mu_i^0$ to the melting point of pure i using equation (17.11) as

$$T_{m_i} = \frac{\dfrac{(u_i^0)^L - (u_i^0)^S}{2}}{k\left[3\ln\left(\dfrac{v_{s_i}}{v_{l_i}}\right) + 1\right]} = \frac{\Delta e_f}{\Delta s_f}. \quad (23.20)$$

A NUMERICAL EXAMPLE

We may evaluate these equations using the same type of model developed in Chapter 17, when we considered one-component phase diagrams. That is, we will assume a nearest-neighbor model, with 6-12-type pair potentials having well depths ε_{ij} and equilibrium interatomic distance $(r_0)_{ij}$.

To save time, we will let component A in this case be the same species that we treated in the numerical example of Chapter 17, for which we specified an atomic weight of 50, a face-centered cubic structure in the solid phase, and

$$\rho_{s_A} = 5.00\,\text{g/cm}^3, \quad \rho_{l_A} = 4.20\,\text{g/cm}^3, \quad \varepsilon_{AA} = 40{,}000\,\text{J/mol}, \quad (23.21)$$

leading to

$$T_{m_A} = 768\,\text{K}, \quad (u_{0_A})^S = -480\,\text{kJ/mol}, \quad (u_{0_A})^L = -438\,\text{kJ/mol},$$
$$v_{s_A} = 1.5 \times 10^{13}\,\text{sec}^{-1}, \quad v_{l_A} = 7.0 \times 10^{12}\,\text{sec}^{-1}. \quad (23.22)$$

In the present case, we will use the same model to describe component B, having an atomic weight of 40, the same crystal structure, and

$$\rho_{s_B} = 5.00\,\text{g/cm}^3, \quad \rho_{l_B} = 4.20\,\text{g/cm}^3, \quad \varepsilon_{BB} = 30{,}000\,\text{J/mol}, \quad (23.23)$$

which leads to

$$T_{m_B} = 568\,\text{K}, \quad (u_{0_B})^S = -360\,\text{kJ/mol}, \quad (u_{0_B})^L = -329\,\text{kJ/mol},$$
$$v_{s_B} = 1.45 \times 10^{13}\,\text{sec}^{-1}, \quad v_{l_B} = 6.78 \times 10^{12}\,\text{sec}^{-1}. \quad (23.24)$$

We may use these values of T_{m_i}, along with the values of $(\Delta S_f)_i$ calculated from

$$(\Delta S_f)_i = R\left[3\ln\left(\frac{v_{s_i}}{v_{l_i}}\right) + 1\right], \quad (23.25)$$

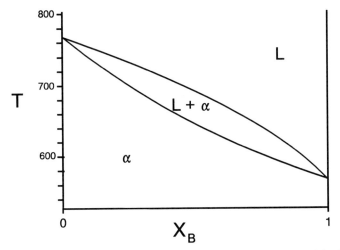

Figure 23.2 The phase diagram for a two-component system in which the solid and liquid phases both show ideal behavior. The interaction potential is assumed to be of the Lennard-Jones 6-12 type.

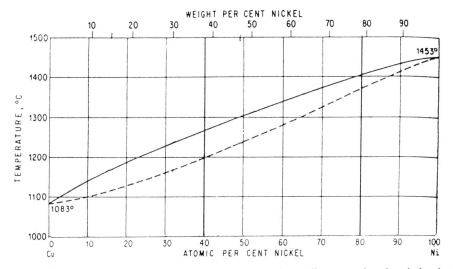

Figure 23.3 The copper–nickel two-component phase diagram, showing behavior close to that predicted for ideal solutions. *Source:* M. Hansen, *Constitution of Binary Alloys*, Second Edition, 1958. Reprinted with permission of McGraw-Hill, Inc.

where we have used equations (14.41) and (16.21) for the entropies of pure solid and liquid respectively, to evaluate equations (23.17) over a range of temperatures, and thus construct the desired phase diagram for this ideal system. The calculated values are

$$(\Delta S_f)_A = 27.3 \text{ J/mol-K}, \qquad (\Delta S_f)_B = 26.1 \text{ J/mol-K}. \qquad (23.26)$$

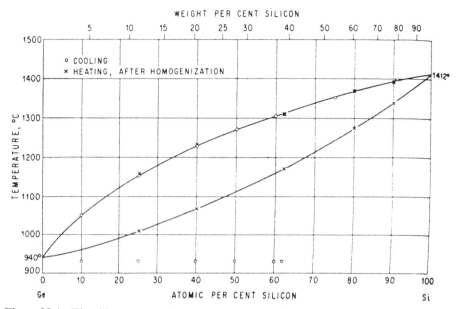

Figure 23.4 The silicon–germanium two-component phase diagram, showing behavior close to ideal, but with larger ΔS_f terms than the copper–nickel system. *Source:* M. Hansen, *Constitution of Binary Alloys*, Second Edition, 1958. Reprinted with permission of McGraw-Hill, Inc.

A plot of the resulting values for X_B^S and X_B^L versus T is shown in Figure 23.2. This result may be compared to the observed phase diagrams for the copper–nickel and germanium–silicon systems, shown in Figures 23.3 and 23.4, respectively. The behavior of these two systems is close to ideal. Note that the germanium–silicon system shows a much wider "loop" than the copper–nickel system. This is a consequence of the much larger ΔS_f for the germanium–silicon system.

GAS-PHASE–CONDENSED-PHASE EQUILIBRIUM

To complete our discussion of two-component ideal systems, let us consider the equilibrium vapor pressures of A and B over the condensed phases. To do this, we return to the basic equilibrium expression

$$\mu_i^S = \mu_i^L = \mu_i^V \qquad (23.27)$$

for the case where the solid and liquid are in equilibrium with the gas phase.

For the case of ideal solid and liquid solutions considered here, this leads to

$$\mu_i^V = \mu_i^S$$

$$-kT \ln\left[\left(\frac{2\pi mkT}{h^2}\right)^{3/2} \frac{kT}{P_i}\right] = \frac{(u_{0_i})^S}{2} - 3kT \ln\left(\frac{kT}{hv_{s_i}}\right) + kT \ln X_i, \qquad (23.28)$$

for the equilibrium between gas and solid, where we have assumed that the gas phase is monatomic and the temperature is high enough that the vibrational modes of the solid behave classically. This equation may be solved for P_i at any temperature if m, u_{0_B} and v_{s_i} are known. For case of gas–liquid equilibrium, one would have to use the corresponding equation for μ_i^L.

TWO-COMPONENT NONIDEAL SYSTEMS

In order to produce commonly observed phase diagram features such as phase separation, melting point maxima or minima, and eutectics or peritectics, we must consider the behavior of nonideal systems. We will do this for the case of a system which has (a) a single liquid phase whose behavior is ideal and (b) a solid solution whose behavior follows the regular, or Bragg–Williams, solution model. Here again these restrictions are made only for mathematical convenience.

For this case, again using the formulations developed in Chapter 22, we have

$$F^L = X_A^L N_{Av}(\mu_A^0)^L + X_B^L N_{Av}(\mu_B^0)^L + RT(X_A^L \ln X_A^L + X_B^L \ln X_B^L),$$

$$F^S = X_A^S N_{Av}(\mu_A^0)^S + X_B^S N_{Av}(\mu_B^0)^S + RT(X_A^S \ln X_A^S + X_B^S \ln X_B^S) - \frac{cw}{2} X_A^S X_B^S,$$

$$(23.29)$$

where

$$w = 2\varepsilon_{AB} - \varepsilon_{AA} - \varepsilon_{BB} \qquad (23.30)$$

as before and may be either greater than or less than zero.

PHASE SEPARATION

Let us begin consideration of the phase diagram features associated with nonideal solutions by considering in more detail the behavior of the nonideal solid phase. As was shown in Chapter 22, for the case where w is less than zero, the free energy versus composition curves become double valued at low temperatures, implying separation into two phases, α and α'. (The relevant curves are shown in Figure 22.3.) We may determine the location of the phase-separated region on the phase diagram by realizing that, at temperatures below the critical temperature for phase separation,

$$\mu_B^\alpha = \mu_B^{\alpha'}$$

$$(\mu_B^0)^\alpha + kT \ln X_B^\alpha - \frac{cw}{2}(1 - X_B^\alpha)^2 = (\mu_B^0)^{\alpha'} + kT \ln X_B^{\alpha'} - \frac{cw}{2}(1 - X_B^{\alpha'})^2.$$

(23.31)

This expression may be simplified if we realize that

$$(\mu_B^0)^\alpha = (\mu_B^0)^{\alpha'},$$ (23.32)

as both refer to pure solid B having the same structure. Furthermore, because the system behavior is symmetrical about $X_B = 0.5$, we may write that

$$X_B^\alpha = 1 - X_B^{\alpha'}.$$ (23.33)

making these substitutions into equation (23.31) leads to

$$kT \ln \frac{X_B^\alpha}{1 - X_B^\alpha} - \frac{cw}{2}(1 - 2X_B^\alpha) = 0.$$ (23.34)

This may be solved for T to yield

$$T = \frac{\frac{cw}{2k}(1 - 2X_B^\alpha)}{\ln \frac{X_B^\alpha}{1 - X_B^\alpha}}.$$ (23.35)

A NUMERICAL EXAMPLE

In order to plot equation (23.35) for the model that we have been using, we must be able to evaluate w. To do this, we must have a pair potential for the interaction between an A atom and a B atom. We will retain the values for AA and BB pair potentials used in the ideal solution case and introduce

$$\varepsilon_{AB} = 34{,}500 \text{ J/mol}.$$ (23.36)

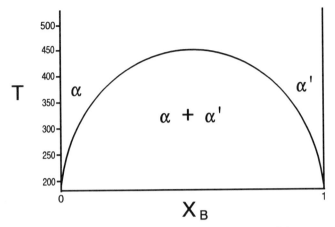

Figure 23.5 The two-component phase diagram for the solid-state portion of a system in which phase separation occurs in the solid phase due to a net repulsive interaction between A and B atoms on adjacent sites, according to the Bragg–Williams model. The interaction potential is assumed to be of the Lennard-Jones 6-12 type.

This leads to

$$w = 2\varepsilon_{AB} - \varepsilon_{AA} - \varepsilon_{BB} = -1250 \text{ J/mol},$$

$$\frac{cw}{2} = -7500 \text{ J/mol}. \tag{23.37}$$

The resulting plot of equation (23.35) is shown in Figure 23.5. As expected, the figure shows phase separation for all temperatures below $T_c = -cw/4k$, with the equilibrium concentrations in the two phases approaching zero at very low temperature.

SOLID–LIQUID EQUILIBRIUM IN NONIDEAL SYSTEMS

Let us again consider the case in which w is negative (like pairs favored). In this case, the behavior of the free energy versus composition curves at various temperatures is shown schematically in Figure 23.6. At temperatures above the melting point of pure A, the solid curve will lie above that of the liquid at all values of X_B, with the liquid being the stable phase. At very low temperatures the free energy curve for the solid will lie below that of the liquid at all X_B, and will be double-valued at some temperatures, leading to phase separation, as we have seen above. At some intermediate temperatures the two curves will cross and there will be a range of solid–liquid equilibrium.

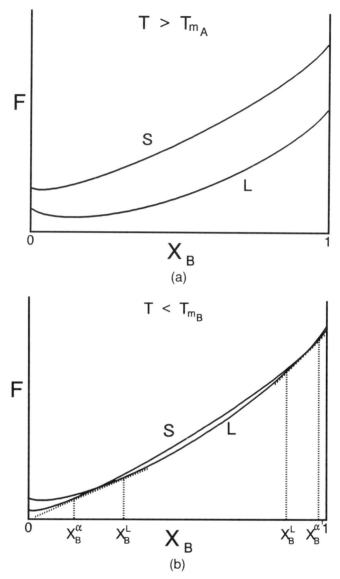

Figure 23.6 Free energy versus composition curves for the solid and liquid phases in a two-component system, with $T_{m_A} > T_{m_B}$, in which the liquid phase is ideal, but the solid phase shows nonideal behavior, with a net repulsive interaction between A and B atoms on adjacent sites. (a) $T_1 > T_{m_A}$. (b) $T_2 < T_{m_B}$. (c) $T_3 \ll T_{m_B}$. At T_2, a liquid and two solid phases are present, with compositions determined by the common tangents to the liquid and solid phase curves. At T_3, two solid phases are present, with compositions determined by the common tangent to the two minima in the solid-phase free energy curve.

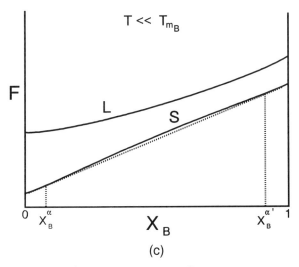

Figure 23.6 (*Continued*).

We may determine the equilibrium compositions of the solid and liquid phases in this case as we did for the ideal solution case above, by equating the chemical potentials of the various possible phases. In this case we may write, using the models of Chapter 22,

$$\mu_i^L = (\mu_i^0)^L + kT \ln X_i^L,$$

$$\mu_i^S = (\mu_i^0)^S + kT \ln X_i^S - \frac{cw}{2}(1 - X_i^S)^2. \tag{23.38}$$

At equilibrium, we have again

$$\mu_A^L = \mu_A^S,$$
$$\mu_B^L = \mu_B^S, \tag{23.39}$$

which, in this case, leads to

$$kT \ln \frac{X_B^S}{X_B^L} - \frac{cw}{2}(1 - X_B^S)^2 = (\Delta\mu^0)_B,$$

$$kT \ln \frac{1 - X_B^S}{1 - X_B^L} - \frac{cw}{2}(X_B^S)^2 = (\Delta\mu^0)_A. \tag{23.40}$$

Again, we have two equations in the unknowns X_B^S and X_B^L. In this case, however, the equations are transcendental and cannot be solved analytically. We must thus proceed directly to a numerical solution, using the appropriate values of the $(\mu_0)_i^\alpha$, c, and w.

296 PHASE EQUILIBRIUM IN MULTICOMPONENT SYSTEMS

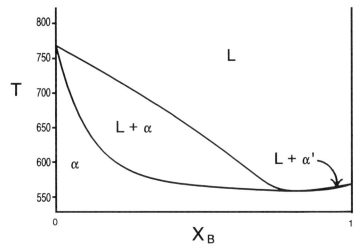

Figure 23.7 The two-component phase diagram for solid–liquid equilibrium a system in which the liquid phase is ideal and the solid phase shows nonideal behavior, with a net repulsive interaction between A and B atoms on adjacent sites, according to the Bragg–Williams model. The interaction potential is assumed to be of the Lennard-Jones 6-12 type.

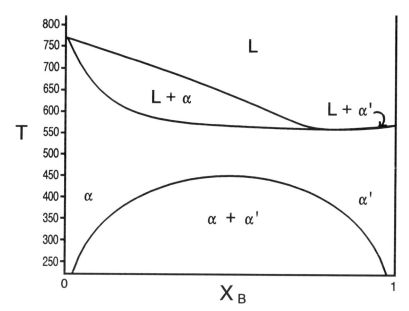

Figure 23.8 The complete two-component phase diagram for the system treated in figures 23.5 and 23.7, showing both solid–liquid equilibrium and phase separation in the solid phase.

A NUMERICAL EXAMPLE

The results of the numerical solution of equation (23.40) are shown in Figure 23.7. Here, rather than the steady decrease in solidus and liquidus temperatures with increasing X_B seen in the ideal solution case, there is a coincident minimum in the two curves, leading to a melting point minimum. We will not

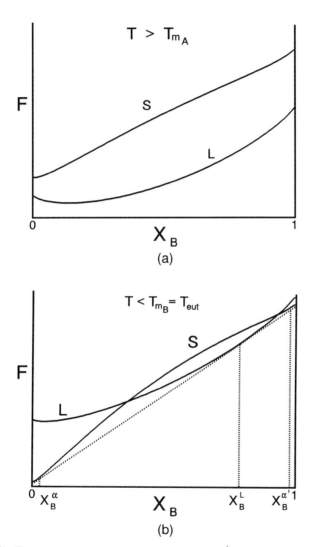

Figure 23.9 Free energy versus composition curves for a two-component system in which the liquid phase is ideal, but the solid phase shows nonideal behavior, with a net repulsive interaction between A and B atoms on adjacent sites, with a larger repulsive interaction energy than in Figure 23.6. (a) $T_1 > T_{m_A}$, T_{m_B}. (b) $T_2 < T_{m_B} = T_{eut}$. (c) $T_3 \gg T_{m_A}$, T_{m_B}. *(Continued on next page)*

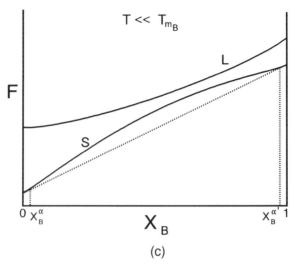

(c)

Figure 23.9 (*Continued*).

consider the case of *w* greater than zero in detail, but merely state that this would lead to a melting point maximum. The complete phase diagram for this example, including both the solid–liquid equilibrium and the phase separation in the solid phase, is shown in Figure 23.8.

As a final example in this kind of system, in which we have an ideal liquid phase and a nonideal solid phase, with a net repulsive interaction between *A* and *B* atoms, we may consider a case in which the value of *w* is more negative than in the previous case. In this final case we will use

$$\frac{cw}{2} = -16{,}500 \text{ J/mol}. \tag{23.41}$$

The resulting free energy versus composition curves are shown in Figure 23.9. Again, at temperatures above the melting point of pure *A*, the solid curve will lie completely above the liquid curve, with the liquid phase being stable at all compositions. Again, at low temperatures, the solid curve will lie completely below the liquid curve and will be double valued, leading to phase separation. At some temperature below the melting point of pure *B*, it will be possible to construct a single tangent line connecting both minima on the solid curve and the minimum in the liquid curve. This gives use to a new feature on the phase diagram, namely a eutectic point. The phase diagram for this case may be obtained by numerical solution of equations (23.34) and (23.40), and is shown in Figure 23.10. It shows all of the features commonly observed in eutectic systems.

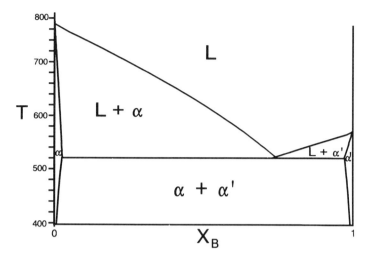

Figure 23.10 The two-component phase diagram for a system in which the liquid phase is ideal but the solid phase shows nonideal behaviour, with a large net repulsive interaction between A and B atoms on adjacent sites, according to the Bragg–Williams model, with the Lennard-Jones 6-12 interaction potential, showing a eutectic reaction.

We will not carry the argument here any further, although we could apply the principles introduced here to systems containing any combination of ideal or nonideal solid and liquid phases. In all cases, the basic criterion for equilibrium will be given by equation (23.1).

BIBLIOGRAPHY

R. T. DeHoff, *Thermodynamics in Materials Science*, McGraw-Hill, New York, 1993, Chapters 9 and 10.

D. R. Gaskill, *Introduction to Metallurgical Thermodynamics*, McGraw-Hill, New York, 1973, Chapter 12.

P. Gordon, *Principles of Phase Diagrams in Materials Systems*, McGraw-Hill, New York, 1968, Chapters 4–8.

C. C. Kittel and H. Kroemer, *Thermal Physics*, 2nd ed., W. H. Freeman & Co., San Francisco, 1980, Chapter 11.

J. W. Whalen, *Molecular Thermodynamics*, John Wiley & Sons, New York, 1991, Chapter 10.

PROBLEMS

23.1 The following information is available for a system consisting of a mixture of Ar and Kr atoms: The molecules Ar_2, Kr_2, and $ArKr$ are unstable in the gas phase. Mixtures of Ar and Kr form random solid solutions having the face-centered cubic crystal structure. Theoretical calculations of interatomic interactions in the solid state yielded the following:

	Ar	Kr
	$M = 40$ amu	$M = 85$ amu
	$\varepsilon_{Ar-Ar} = -3000$ J/mol	$\varepsilon_{Kr-Kr} = -5800$ J/mol
	$\Theta_{E_{Ar}} = 40$ K	$\Theta_{E_{Kr}} = 60$ K
	$\varepsilon_{Ar-Kr} = -4600$ J/mol.	

For a mixture of 10% Ar, 90% Kr at 20 K:

(a) What is the activity of Kr in the solid phase?

(b) What is the equilibrium partial pressure of Kr over the solid phase?

23.2 Two metals A and B form regular solid solutions with $cw/2 = -5000$ J/mol and ideal liquid solutions. The melting point of A is 900 K and its entropy of fusion is 15 J/mol-K. The melting point of B is 1000 K and its entropy of fusion is 18 J/mol-K.

(a) Will minima occur in the solidus and liquidus curves?

(b) If so, calculate the temperature of the minimum.

23.3 Solid solutions of species A and B are regular in their thermodynamic behavior and have $cw/2 = -1500$ J/mol for the 50–50 atom percent mixture. Liquid solutions of these two species are ideal. The equilibrium vapor phase over these alloys consists of monatomic A and B atoms. The relevant physical properties of A and B are:

Property	A	B
Atomic mass (amu)	200	150
Crystal structure	FCC	FCC
Molar volume (cm³/mol)	10 (xtal)	10 (xtal)
	11 (liq)	10.5 (liq)
Lattice parameter (Å)	3.00	3.00
Melting point (K)	1000	900
Boiling point (K)	2500	2200
Heat of fusion (J/mol)	12,000	10,000
Heat of vaporization (J/mol)	240,000	160,000

(a) Plot the Helmholz free energy of the solid and liquid alloys versus composition for $T = 800$ K.

(b) Determine the equilibrium compositions of all phases present at 800 K.

23.4 Draw appropriately labeled free energy versus composition diagrams at the indicated temperatures for the two binary systems whose phase diagrams appear below.

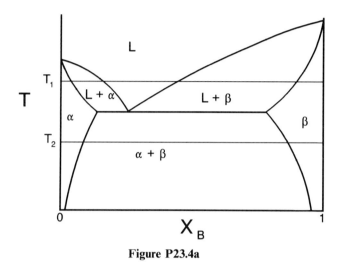

Figure P23.4a

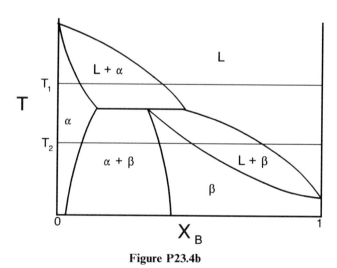

Figure P23.4b

23.5 Atomic species A and B form regular solid solutions which are FCC crystals. $(\Delta H_s)_A = 60{,}000$ cal/mol, $(\Delta H_s)_B = 40{,}000$ cal/mol. The strength of an AB bond is 0.9 times that of the average strength of AA and BB bonds. A and B have monatomic vapor phases, $M_A = 54$, $M_B = 40$, and electronic excitation of A and B in the vapor phase may be neglected. At $T = 800$ K, what are the compositions of all phases present at equilibrium for an overall composition of 40% B? (Neglect the possible presence of a liquid phase.)

23.6 Elements A and B form ideal solutions in both the solid and liquid states. The properties of the system may be modeled in terms of nearest-neighbor model, with the Lennard-Jones 6-12 potential. The appropriate parameters are:

Parameter	A	B
c	12	12
ε (J/mol)	30,000	40,000
$r_e = r_{0_S}$ (Å)	3.00	3.00
r_{0_L} (Å)	3.15	3.15
M (amu)	40	60

What are the compositions of all phases present at equilibrium for $X_B = 0.5$ and a temperature halfway between the melting points of pure A and pure B?

CHAPTER 24

CHEMICAL EQUILIBRIUM

We turn now to the application of statistical thermodynamics to the determination of chemical equilibrium. We will do this using the general equation for chemical equilibrium developed in Chapter 5, namely

$$\sum_i v_i \mu_i = \sum_j v_j \mu_j, \qquad (24.1)$$

in which the μ_i are the chemical potentials of the various species consumed in the reaction and the v_i are their corresponding stoichiometric coefficients and the μ_j and v_j are the corresponding terms for the species formed in the reaction. In all cases, we will use the expressions for the various μ_i and μ_j developed in our previous treatments of solids, liquids, and gases.

THE EQUILIBRIUM CONSTANT

Let us begin by considering the free energy change accompanying a chemical reaction such as

$$aA + bB \to cC + dD, \qquad (24.2)$$

where the A, B, C, and D may be any chemical species in either a condensed or gas phase. In general, we may write the following equation for the reaction of v_A moles of A plus v_B moles of B to form v_C moles of C and v_D moles of D:

$$\Delta G = -v_A \mu_A N_{Av} - v_B \mu_B N_{Av} + v_C \mu_C N_{Av} + v_D \mu_D N_{Av}. \qquad (24.3)$$

At equilibrium, under conditions of constant T and P we have

$$\Delta G = 0, \quad (24.4)$$

and thus

$$\nu_A \mu_A + \nu_B \mu_B = \nu_C \mu_C + \nu_D \mu_D, \quad (24.5)$$

which is the same as equation (24.1).

It is often convenient to rewrite this equation using the general relation

$$\mu_i = \mu_i^0 + kT \ln a_i. \quad (24.6)$$

This yields, after some rearrangement,

$$\nu_A \mu_A^0 + \nu_B \mu_B^0 - \nu_C \mu_C^0 - \nu_D \mu_D^0 = kT \ln \left(\frac{a_C^{\nu_C} a_D^{\nu_D}}{a_A^{\nu_A} a_B^{\nu_B}} \right)_{equil}. \quad (24.7)$$

Or, if we define ΔG^0, the standard free energy change for the reaction, as

$$\Delta G^0 = N_{Av}(\nu_C \mu_C^0 + \nu_D \mu_D^0 - \nu_A \mu_A^0 - \nu_B \mu_B^0) \quad (24.8)$$

and we further define K_a, the equilibrium constant for the reaction expressed in terms of activities, as

$$K_a \equiv \left(\frac{\prod_j a_j^{\nu_j}}{\prod_i a_i^{\nu_i}} \right)_{equil}, \quad (24.9)$$

in which the products are taken over all of the i reactant species and all of the j product species, we have as the final relation defining chemical equilibrium

$$\Delta G^0 = -RT \ln K_a. \quad (24.10)$$

Note that the value of ΔG^0 will depend on the standard states chosen for the various reactants and products and represents the change in free energy associated with the reaction at the chosen standard state.

DISSOCIATION OF A DIATOMIC MOLECULE

The simplest case of chemical reaction to deal with mathematically is the dissociation of a homonuclear diatomic molecule according to the reaction

$$A_2 \leftrightarrows 2A. \quad (24.11)$$

For this reaction, at equilibrium,

$$\mu_{A_2} = 2\mu_A, \tag{24.12}$$

or

$$K_a = \left(\frac{a_A^2}{a_{A_2}}\right). \tag{24.13}$$

We may use the expressions for μ_A and μ_{A_2} developed in Chapters 12 and 13, respectively, namely

$$\mu_A = -kT \ln\left[\left(\frac{2\pi mkT}{h^2}\right)^{3/2} \frac{kT}{P_A}\right], \tag{24.14}$$

and

$$\mu_{A_2} = -kT \ln\left[\left(\frac{2\pi(2m)kT}{h^2}\right)^{3/2} \frac{kT}{P_{A_2}}\right] - kT \ln\frac{T}{\sigma\Theta_r} + \frac{k\Theta_v}{2}$$
$$+ kT \ln(1 - e^{-\Theta_v/T}) - D_e - kT \ln \omega_{e_1}. \tag{24.15}$$

If we substitute these expressions into equation (24.12), we obtain

$$\mu_{A_2}^0(T) + kT \ln\left(\frac{P_{A_2}}{P^0}\right) = 2\left[\mu_A^0(T) + kT \ln\left(\frac{P_A}{P^0}\right)\right], \tag{24.16}$$

where

$$\mu_{A_2}^0(T) = -kT \ln\left[\left(\frac{2\pi(2m)kT}{h^2}\right)^{3/2} \frac{kT}{P^0}\right] - kT \ln\frac{T}{\sigma\Theta_r} + \frac{k\Theta_v}{2}$$
$$+ kT \ln(1 - e^{-\Theta_v/T}) - D_e - kT \ln \omega_{e_1},$$

$$\mu_A^0(T) = -kT \ln\left[\left(\frac{2\pi mkT}{h^2}\right)^{3/2} \frac{kT}{P^0}\right], \tag{24.17}$$

where P^0 is the arbitrarily chosen standard state pressure, in this case usually the system total pressure. We may rearrange Equation 24.16 to obtain

$$2\mu_A^0(T) - \mu_{A_2}^0(T) = -kT \ln \frac{\left(\frac{P_A}{P^0}\right)^2}{\left(\frac{P_{A_2}}{P^0}\right)}. \tag{24.18}$$

Comparison with equations (24.8), (24.9), and (24.10) yields

$$\Delta G^0(T) = N_{Av}[2\mu_A^0(T) - \mu_{A_2}^0(T)],$$

$$K_a = \frac{\left(\dfrac{P_A}{P^0}\right)^2}{\left(\dfrac{P_{A_2}}{P^0}\right)} = \frac{P_A^2}{P_{A_2}P^0}. \tag{24.19}$$

Note that very often equilibria in gas phase reactions are expressed in terms of K_P, the equilibrium constant expressed in terms of the partial pressures of reactants and products. In this case,

$$K_P = \frac{P_A^2}{P_{A_2}} = P_0 K_a. \tag{24.20}$$

The equilibrium degree of dissociation at any temperature and total pressure may thus be found from equation (24.18) and the additional condition that

$$P^0 = P_{A_2} + P_A, \tag{24.21}$$

if the appropriate values of m, Θ_r, Θ_v, D_e, and ω_{e_1} are known. The equilibrium constant, K_a, may be evaluated by using

$$\frac{\Delta G^0}{N_{Av}} = 2\mu_A^0 - \mu_{A_2}^0, \tag{24.22}$$

which, after inserting the expressions from equation (24.17) for the μ_i^0 terms and some rearrangement, yields

$$\frac{\Delta G^0}{RT} = -\ln\left[\left(\frac{2\pi kT}{h^2}\right)^{3/2} \frac{kT}{P^0} \left(\frac{m^2}{2m}\right)^{3/2} \left(\frac{\sigma\Theta_r}{T}\right) \left(\frac{1 - e^{-\Theta_v/T}}{e^{\Theta_v/2T}}\right) \frac{e^{-D_e/kT}}{\omega_{e_1}}\right] \tag{24.23}$$

or

$$K_a = \frac{1}{P^0}\left[\left(\frac{2\pi kT}{h^2}\right)^{3/2} kT \left(\frac{m}{2}\right)^{3/2} \left(\frac{\sigma\Theta_r}{T}\right) \left(\frac{1 - e^{-\Theta_v/T}}{e^{-\Theta_v/2T}}\right) \frac{e^{-D_e/kT}}{\omega_{e_1}}\right]. \tag{24.24}$$

from this we may obtain

$$\frac{P_A^2}{P_{A_2}} = \left[\left(\frac{2\pi kT}{h^2}\right)^{3/2} kT \left(\frac{m}{2}\right)^{3/2} \left(\frac{\sigma\Theta_r}{T}\right) \left(\frac{1 - e^{-\Theta_v/T}}{e^{-\Theta_v/2T}}\right) \frac{e^{-D_e/kT}}{\omega_{e_1}}\right], \tag{24.25}$$

the desired ratio of the equilibrium partial pressures. This equation, along with equation (24.21) and the appropriate values of the molecular parameters, permits calculation of the desired partial pressures.

ISOTOPIC EQUILIBRIUM

The next case that we will consider is the equilibrium among the various isotopes of a diatomic gas. We will consider the reaction

$$A_2 + B_2 \rightleftarrows 2AB, \tag{24.26}$$

in which A and B are two isotopes of the same chemical species—for example, hydrogen and deuterium, C^{12} and C^{13}, or O^{16} and O^{18}.

For this case we have, at equilibrium,

$$\mu_{A_2} + \mu_{B_2} = 2\mu_{AB}, \tag{24.27}$$

or

$$K_a = \left(\frac{a_{AB}^2}{a_{B_2} a_{A_2}}\right)_{equil} = \frac{P_{AA}^2}{P_{AA} P_{BB}}. \tag{24.28}$$

In this case, since all species involved are diatomic molecules, we have

$$\mu_i = -kT \ln\left[\left(\frac{2\pi m_i kT}{h^2}\right)^{3/2} \frac{kT}{P_i}\right] - kT \ln \frac{T}{\sigma_i \Theta_{r_i}} + \frac{k\Theta_{v_i}}{2}$$
$$+ kT \ln(1 - e^{-\Theta_{v_i}/kT}) - D_{e_i} - kT \ln \omega_{e_{1_i}}. \tag{24.29}$$

For this case of isotopic equilibrium, D_e, r_e, the equilibrium interatomic spacing, and ω_{e_1} will have essentially the same values for all of the species involved, because the shape of the potential well is determined primarily by electron–electron interactions, and the electronic configuration is essentially the same for all isotopes of a given material. Consequently all terms involving D_e and ω_e will vanish from the final expression for K_a.

In addition, since there is no change in the total number of moles of gas present, all terms in the translational contribution to K_a will vanish except for the terms in m_{ii} and P_{ii}. We may thus write that at equilibrium

$$\mu_{AA} + \mu_{BB} - 2\mu_{AB} = 0. \tag{24.30}$$

Substituting equation (24.29) into equation (24.30) and eliminating common factors yields

$$\ln\left[\left(\frac{m_{AA}m_{BB}}{m_{AB}}\right)^{3/2}\frac{P_{AB}^2}{P_{AA}P_{BB}}\right]+\ln\left(\frac{\sigma_{AB}^2\Theta_{rAB}^2}{\sigma_{AA}\Theta_{rAA}\sigma_{BB}\Theta_{rBB}}\right)+\frac{1}{2T}(2\Theta_{v_{AB}}-\Theta_{v_{AA}}-\Theta_{v_{BB}})$$
$$-\ln\left[\frac{(1-e^{-\Theta_{v_{AA}}/T})(1-e^{-\Theta_{v_{BB}}/T})}{(1-e^{-\Theta_{v_{AB}}/T})^2}\right]=0. \quad (24.31)$$

This may be rearranged to yield

$$K_a=\frac{P_{AB}^2}{P_{AA}P_{BB}}=\left(\frac{m_{AB}^2}{m_{AA}m_{BB}}\right)^{3/2}\left(4\frac{\Theta_{rAA}\Theta_{rBB}}{\Theta_{rAB}^2}\right)e^{-(1/2T)(2\Theta_{v_{AB}}-\Theta_{v_{AA}}-\Theta_{v_{BB}})}$$
$$\cdot\left[\frac{(1-e^{-\Theta_{v_{AA}}/T})(1-e^{-\Theta_{v_{BB}}/T})}{(1-e^{-\Theta_{v_{AB}}/T})^2}\right], \quad (24.32)$$

where we have substituted

$$\sigma_{AA}=\sigma_{BB}=2, \qquad \sigma_{AB}=1. \quad (24.33)$$

We may further simplify this result if we realize that the Θ_r and Θ_v terms for the various species differ only in the dependences of the moments of inertia and the vibrational frequencies on the reduced masses, μ, through

$$I_{ij}=2r_e\mu_{ij}=2r_e\frac{m_im_j}{(m_i+m_j)}$$
$$v_{ij}=\frac{1}{2\pi}\left(\frac{f}{\mu}\right)^{1/2}=\frac{1}{2\pi}\left(\frac{f}{\left(\frac{m_im_j}{m_i+m_j}\right)}\right)^{1/2}. \quad (24.34)$$

Finally, if we introduce the simplifications of Equation (24.34), and assume in addition that the temperature is low enough that we may neglect vibrational excitation, we arrive at

$$K_a=\frac{P_{AB}^2}{P_{AA}P_{BB}}=2\frac{m_A+m_B}{m_A^{1/2}m_B^{1/2}}\exp\left[-\frac{\Theta_{v_{AB}}}{T}\left(1-\frac{m_A^{1/2}+m_B^{1/2}}{2^{1/2}(m_A+m_B)^{1/2}}\right)\right]. \quad (24.35)$$

If one evaluates this expression for the case of hydrogen and deuterium, one determines that $K_a=3.46$ at 383 K. This implies that at equilibrium at this temperature

$$H_2:D_2:HD=1.00:1.00:1.86. \quad (24.36)$$

Note that this result is significantly different than that which would be expected for random mixing of the two isotopes, namely

$$H_2 : D_2 : HD = 1 : 1 : 2. \tag{24.37}$$

Note too, however, that as $m_A/m_B \to 1$, $K_a \to 4$, implying that the random distribution is approached at this limit.

GENERAL GAS-PHASE REACTIONS

As a final case in the study of gas-phase reactions, we will look at the general case, in which different chemical species are involved. The method is the same in this case as in the two cases discussed above; the only differences are that the simplifications that arose in the previous cases may no longer be made, and that, in many cases of practical interest, we must deal with polyatomic molecules. These complications lead to more complicated expressions for K_a, and consequently involve more mathematics in the determination of K_a. However, as long as the basic data on D_e, m's, Θ_r's, and vibrational frequencies for each molecule are available, no new approximations are required. Consider as an example the so-called "water–gas" reaction

$$CO_2 + H_2 \rightleftarrows CO + H_2O. \tag{24.38}$$

For this reaction, in general,

$$K_a = \frac{a_{CO} a_{H_2O}}{a_{CO_2} a_{H_2}} = \frac{P_{CO} P_{H_2O}}{P_{CO_2} P_{H_2}}. \tag{24.39}$$

And, at equilibrium,

$$\mu_{CO} + \mu_{H_2O} - \mu_{CO_2} - \mu_{H_2} = 0, \tag{24.40}$$

The expressions for μ_{H_2} and μ_{CO} will be of the same form as for the diatomic molecules that we have considered previously. CO_2 is a linear triatomic molecule, so that the expression for Θ_r will be same as for the diatomic molecule. However, there must be four vibrational modes to yield the required total of three modes per atom in the molecule. For the water molecule, which is a nonlinear triatomic molecule, we must account for three rotational degrees of freedom and three vibrational degrees of freedom. The appropriate values of the Θ_r and Θ_v are tabulated in Appendix F.

For the case where the temperature is high enough that all rotational terms may be treated classically, and realizing again that in this reaction in which the number of moles of gas does not change, the translational terms will reduce to expressions involving only masses and pressures, and finally that $\omega_{e_1} = 1$ for all of the species involved, we may write for equation (24.40) that

$$\ln\left[\left(\frac{m_{CO}m_{H_2O}}{m_{CO_2}m_{H_2}}\right)^{3/2}\frac{P_{CO_2}P_{H_2}}{P_{CO}P_{H_2O}}\right] + \ln\left[\frac{\prod_R(\sigma_r\Theta_r)}{\prod_P(\sigma_r\Theta_r)}\right] + \frac{k}{2}\left(\sum_P\Theta_v - \sum_R\Theta_v\right)$$

$$+ \ln\left[\frac{\prod_R(1-e^{-\Theta_v/T})}{\prod_P(1-e^{-\Theta_v/T})}\right] - \frac{(D_{e_{CO}} + D_{e_{H_2O}} - D_{e_{CO_2}} - D_{e_{H_2}})}{kT} = 0, \quad (24.41)$$

where the products and sums are taken over all rotational and vibrational modes of products (P) and reactants (R) as indicated. This expression leads to

$$K_a = \frac{P_{CO}P_{H_2O}}{P_{CO_2}P_{H_2}} = \left(\frac{m_{CO}m_{H_2O}}{m_{CO_2}m_{H_2}}\right)^{3/2}\left[\frac{\prod_R(\sigma_r\Theta_r)}{\prod_P(\sigma_r\Theta_r)}\right]\left[\frac{\prod_R(1-e^{-\Theta_v/T})}{\prod_P(1-e^{-\Theta_v/T})}\right]e^{\Delta E_0/kT}$$

$$(24.42)$$

where

$$\Delta E_0 = D_{e_{CO}} + D_{e_{H_2O}} - D_{e_{CO_2}} - D_{e_{H_2}} - \frac{k}{2}\left(\sum_P\Theta_v - \sum_R\Theta_v\right), \quad (24.43)$$

and represents the difference in energy between the products and reactants at rest in their ground states. For this reaction, $\Delta E_0 = -40336$ J/mol, as determined from a combination of calorimetric and spectroscopic data.

HETEROPHASE REACTIONS

For cases in which one or more of the species involved in the reaction is a solid at the temperature of the reaction, additional complications arise. Equation (24.1) is still the basic criterion for equilibrium in these cases, but the expression for the chemical potentials of the solid species must be obtained from the appropriate model for the solid.

In the case where the only solid involved is a pure crystalline material that can be treated by the Einstein or Debye models, no significant problems arise. For example, if we consider the gasification of carbon according to the reaction

$$C^S + O_2^G \leftrightarrows CO_2^G, \quad (24.44)$$

we may write that

$$K_a = \frac{a_{CO_2}}{a_C a_{O_2}} = \frac{P_{CO_2}}{P_{O_2}}, \qquad (24.45)$$

because the activity of the pure solid is, by definition, unity. We may determine the equilibrium constant as usual, starting from

$$\mu_{CO_2} - \mu_C - \mu_{O_2} = 0. \qquad (24.46)$$

Substituting the appropriate values for the chemical potentials, using values of u_0 and Θ_E appropriate to graphite, since this is the stable form of carbon at room temperature and atmospheric pressure, leads to

$$\ln\left[\left(\frac{m_{CO_2}}{m_{O_2}}\right)^{3/2} \frac{P_{O_2}}{P_{CO_2}}\right] + \ln\left(\frac{\sigma_{O_2}\Theta_{r_{O_2}}}{\sigma_{CO_2}\Theta_{r_{CO_2}}}\right) + \sum_{\Theta_{v_{CO_2}}} \ln\left(\frac{e^{-\Theta_v/2kT}}{1-e^{-\Theta_v/kT}}\right)$$

$$- \ln\left(\frac{e^{-\Theta_{v_{O_2}}/2kT}}{1-e^{-\Theta_{v_{O_2}}/kT}}\right) + 3\ln\left(2\sinh\left(\frac{\Theta_{E_C}}{2T}\right)\right) + \frac{D_{e_{CO_2}}}{kT}$$

$$- \frac{D_{e_{O_2}}}{kT} + \frac{u_{0_C}}{2kT} + \ln\frac{\omega_{e_{CO_2}}}{\omega_{e_{O_2}}} = 0. \qquad (24.47)$$

This expression may be rearranged to yield the equilibrium constant as

$$K_a = \frac{P_{CO_2}}{P_{O_2}} = \left(\frac{m_{CO_2}}{m_{O_2}}\right)^{3/2} \left(\frac{\sigma_{O_2}\Theta_{r_{O_2}}}{\sigma_{CO_2}\Theta_{r_{CO_2}}}\right) \left[\prod_{\Theta_{v_{CO_2}}} \left(\frac{e^{-\Theta_v/2kT}}{(1-e^{-\Theta_v/T})}\right)\right]$$

$$\cdot \left(\frac{1-e^{-\Theta_{v_{O_2}}/T}}{e^{-\Theta_{v_{O_2}}/T}}\right) \left[2\sinh\left(\frac{\Theta_{E_C}}{2T}\right)\right]^3 \left(\frac{\omega_{e_{CO_2}}}{\omega_{e_{O_2}}}\right) \exp\left(\frac{D_{e_{CO_2}} - D_{e_{O_2}} + u_{0_C}/2}{kT}\right). \qquad (24.48)$$

In principle, this expression can be evaluated at any temperature if all of the molecular level parameters are known or can be calculated. In practice, it is customary to begin with the standard free energy of the reaction, ΔG^0, defined in equation (24.9), which is simply the Gibbs free energy change per mole for the reaction at the chosen standard pressure, usually one atmosphere, at any temperature. For the reaction considered here, the right-hand side of equation (24.48) is simply

$$e^{-\Delta G^0/RT}. \qquad (24.49)$$

This parameter has been determined by experiment, or by a combination of calculation and experiment, for a wide variety of reactions at atmospheric pressure over a range of temperatures, and tabulated values are readily available. A list showing ΔG^0 as a function of temperature for several reactions is given in Appendix I.

The concept of a standard free energy of reaction is also useful in determining the equilibrium constant for reactions involving more than one condensed phase species. Consider, for example, the oxidation of a metal to produce a solid oxide, as in

$$M + O_2 \rightleftarrows MO_2, \qquad (24.50)$$

where M can be any divalent metal. We may write the expression for the equilibrium constant for this case as

$$K_a = \frac{a_{MO_2}}{a_M a_{O_2}} = \frac{1}{a_{O_2}} = \frac{P^0_{O_2}}{P_{O_2}}, \qquad (24.51)$$

because the activities of the two pure solid phases are unity. We may, as usual, write an expression for the equilibrium constant in terms of the chemical potentials of the various species involved as

$$\mu_{MO_2} - \mu_{O_2} - \mu_M = 0, \qquad (24.52)$$

which for this case, assuming that the solids involved can be represented by the Einstein model, leads to

$$\frac{u_{0_{MO_2}}}{2} + 3kT \ln\left(2\sinh\left(\frac{\Theta_{E_{MO_2}}}{2T}\right)\right) - \frac{u_{0_M}}{2} - 3kT \ln\left(2\sinh\left(\frac{\Theta_{E_M}}{2T}\right)\right)$$
$$+ kT \ln\left[\left(\frac{2\pi mkT}{h^2}\right)^{3/2}\frac{kT}{P_{O_2}}\right] + kT \ln\left(\frac{T}{\sigma\Theta_r}\right) - \frac{k\Theta_v}{2}$$
$$- kT \ln(1 - e^{-\Theta_v/T}) + D_e + kT \ln \omega_{e_1} = 0. \qquad (24.53)$$

As in all previous cases treated in this chapter, this expression can be rearranged to yield the equilibrium constant as

$$K_A = \frac{P^0_{O_2}}{P_{O_2}} = \left[\left(\frac{h^2}{2\pi mkT}\right)^{3/2}\frac{P^0_{O_2}}{\omega_{e_1}kT}\right]\left(\frac{T}{\sigma\Theta_r}\right)(1 - e^{-\Theta_v/T})$$
$$\cdot \left(\frac{2\sinh\left(\frac{\Theta_{E_M}}{T}\right)}{2\sinh\left(\frac{\Theta_{E_{MO_2}}}{T}\right)}\right)^3 \exp\left[-\frac{1}{kT}\left(\frac{u_{0_{MO_2}}}{2} - \frac{u_{0_M}}{2} + D_{e_{O_2}} + \frac{\Theta_v}{2}\right)\right].$$
$$(24.54)$$

Again, the right-hand side of this equation is directly related to ΔG^0 for the reaction; this can be obtained, for the case of atmospheric pressure, from the data in Appendix I. The value of the equilibrium constant for any other set of pressure and temperature conditions can be found by using equation (24.54) to calculate the differences in the translational, rotational, and vibrational terms between the standard state and the desired state.

BIBLIOGRAPHY

E. A. Guggenheim, *Thermodynamics*, North Holland, 1957, Chapter 7.

T. L. Hill, *Introduction to Statistical Thermodynamics*, Addison-Wesley, 1960, Chapter 10.

C. C. Kittel and H. Kroemer, *Thermal Physics*, 2nd ed., W. H. Freeman & Co., San Francisco, 1980, Chapter 9.

J. H. Knox, *Molecular Thermodynamics*, John Wiley & Sons, New York, 1978, Chapter 11.

J. W. Whalen, *Molecular Thermodynamics*, John Wiley & Sons, New York, 1991, Chapters 9 and 10.

PROBLEMS

24.1 Calculate the change in Gibbs free energy accompanying the reaction of two moles of ^{16}O atoms to form one mole of $^{16}O_2$ molecules at a constant temperature of 300 K and a constant pressure of 1 atm. Assume that both species are ideal gases. The required energy and characteristic temperature data may be found in Appendix F. The electronic state degeneracies and excited state energies are:

^{16}O	$^{16}O_2$
$\omega_{e_1} = 1$	$\omega_{e_1} = 3$
	$\omega_{e_2} = 1$
	$\varepsilon_{e_2} = 3.70\,\text{eV}$.

24.2 Calculate, using statistical thermodynamics, the equilibrium composition of a system containing one mole of ^{16}O atoms in a one-liter volume at $T = 2500$ K.

24.3 Consider a system consisting, initially, of one mole of $^{35}Cl_2$ molecules in a 10-liter container at 3000 K.

(a) What is the composition of the system if the reaction

$$^{35}Cl_2 \rightleftharpoons 2\,^{35}Cl$$

is allowed to proceed to equilibrium?

(b) What is the change in the Helmholz free energy accompanying this reaction?

For this system we have the following data:

$^{35}\mathrm{Cl}$	$^{35}\mathrm{Cl}_2$
$\omega_{e_1} = 1$	$\omega_{e_1} = 1$
$\omega_{e_2} = 1$	$\omega_{e_2} = 3$
$\varepsilon_{e_2} = 0.35 \text{ eV}$	$\varepsilon_{e_2} = 2.275 \text{ eV}$

24.4 Consider a system consisting of one mole of N_2 molecules in a 10-liter container at 300 K.

(a) What is the change in the Gibbs free energy when the system is heated to 2000 K, assuming that no dissociation takes place?

(b) If a catalyst is added to the system and the reaction

$$N_2 \rightleftharpoons 2N$$

goes to equilibrium, what is the final ratio of N/N_2?

24.5 Consider the chemical reaction

$$N_2(g) + O_2(g) \rightleftharpoons 2NO(g)$$

(a) Develop an expression for K_p for this reaction.

(b) Calculate the equilibrium concentration of NO in air at 25°C.

PART V

QUANTUM SYSTEMS

CHAPTER 25

THE PERFECT ELECTRON GAS

We will consider next a treatment that provides a good approximation to the properties of the free, or nearly free, conduction electrons in the bulk of a metal. We will see that this treatment will have some similarities with that developed for the monatomic ideal gas, but will differ in that the small mass of the electron, along with the high number density of free electrons in a typical solid metal, leads to a system in which the criterion for the use of Boltzmann statistics, namely that the number of available quantum states is large compared to the number of particles, is not met. Consequently, we will have to take explicit account of the fact that electrons, which have half integral spin, follow Fermi–Dirac statistics.

THE MODEL

The model in this case is a solid metal crystal, of volume V, containing positive ions fixed to lattice sites, plus enough free electrons to achieve overall electrical neutrality. The positive charge associated with the ion cores is assumed to be "smeared out" uniformly throughout the volume of the metal, so that the electrons encounter a uniform potential everywhere within the metal. (This model is often referred to as the "jellium" model of the free-electron metal). The electrons are assumed to be free to move anywhere within the crystal. It is assumed that the positive charge associated with the ions serves only to provide a potential well that confines the electrons and that, aside from this, there are no interactions between the ions and the electrons. The potential energy versus distance seen by an electron is thus as shown in Figure 25.1. By

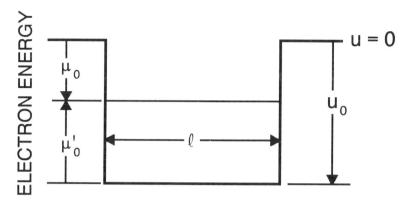

Figure 25.1 Potential well seen by a free electron in a solid described by the "jellium" model.

convention, and to be consistent with our previous definition of the zero of energy, we assume that an electron at rest, far from the crystal and from other electrons, has zero energy. Thus the energy u_0 shown in Figure 25.1 is inherently negative.

The number of free electrons per unit volume will depend on the type of metal considered, but will be in the range of one to four per atom, and is in many cases equal to the number of valence electrons in the free atom. In a more exact treatment, we would see that the electrons are not completely "free" to move in the crystal, because some quantum states are forbidden as a result of electron–ion interactions. For present purposes, we will ignore these effects.

NUMBER OF AVAILABLE STATES

Early treatments of the free electron gas using the model described above did not take account of the fact that, according to quantum mechanics, electrons are fermions (spin = $\frac{1}{2}$) and consequently follow Fermi–Dirac statistics. These treatments led to the erroneous prediction that the free electrons would contribute $\frac{3}{2}$ R per mole to the heat capacity, C_V, whereas the actual contribution is, as we shall see, very much smaller. The key to the correct treatment is the realization that the criterion for the use of Boltzmann statistics, namely that the number of available states be large compared to the number of particles, is not satisfied, as we will now demonstrate.

The kinetic energies available to a given electron in the present case are determined by the same expression as for the ideal monatomic gas. (This again is a case of particles moving in a box, with the potential energy being uniform within the box.) The total energies available are thus

$$\varepsilon_k = u_0 + \frac{h^2}{8m}\left[\left(\frac{k_1}{l_1}\right)^2 + \left(\frac{k_2}{l_2}\right)^2 + \left(\frac{k_3}{l_3}\right)^2\right],$$

$$= u_0 + \frac{h^2}{8mV^{2/3}}(k_1^2 + k_2^2 + k_3^2), \quad k_1, k_2, k_3 = 1, 2, \ldots, \quad (25.1)$$

where the term u_0 accounts for the fact that the electrons are in the potential well formed by the positive ions and we have assumed as before that $l_i = V^{1/3}$. There is, in addition, one further complication: There are two distinguishable quantum states for each value of ε_k due to the two allowed spin states of the electron.

We can determine whether the criterion for the use of Boltzmann statistics is met in this case by using the same argument, based on the concept of the number of states in a volume in one octant of a three-dimensional quantum number space, that we used in the case of the monatomic ideal gas. For this case, as before, the number of quantum states having energies less than some particular value ε is given by

$$G(\varepsilon) = (2)\left(\frac{\pi}{6}\right)\left(\frac{8m\varepsilon}{h^2}\right)^{3/2} V, \quad (25.2)$$

which differs from the expression developed for the ideal gas only by the factor of two discussed above. The number of states with $\varepsilon \leq kT$ is thus

$$G(kT) \approx \left(\frac{8mkT}{h^2}\right)^{3/2} V. \quad (25.3)$$

For the present case we will use the electron mass, $m \approx 10^{-27}$ g, $T \approx 300$ K, and $V \approx 10$ cm^3. This leads to

$$G(kT) \approx 10^{20}. \quad (25.4)$$

Since the number of free electrons in this volume of a typical metal is on the order of 10^{24}, we see that the conditions for the use of Boltzmann statistics are clearly not met, and we must use an expression for the partition function that takes explicit account of the fact that electrons follow Fermi–Dirac statistics. A corollary of this result is that, since the number of particles is large compared to the number of states having energies less than or equal to kT, then there must be many electrons in the system with translational energies large compared to kT, even at very low temperatures. We will see the consequences of this later.

EVALUATION OF THE PARTITION FUNCTION

To treat systems of this sort, it is most convenient to use the grand canonical ensemble partition function,

$$\Xi = \sum_i e^{-\beta E_i} e^{-\gamma N_i}, \tag{25.5}$$

as the number of particles per state enters explicitly in this formulation.

Since we are dealing with a system of identical, indistinguishable particles, with negligible effects of interparticle forces, we can write the following equations for any state of the macroscopic system:

$$E_i = \sum_k \varepsilon_k N_{k_i}, \tag{25.6}$$

$$N_i = \sum_k N_{k_i}, \tag{25.7}$$

where ε_k is the energy of the kth single particle quantum state and N_{k_i} is the number of particles in this *single-particle* state when the system is in a *system* quantum state i. This leads to

$$\Xi = \sum_i \prod_k e^{-[(\beta \varepsilon_k + \gamma) N_{k_i}]}. \tag{25.8}$$

One can show, by reasoning that we will not go into, that it is possible to rewrite this expression as a product of sums, rather than a sum of products, to yield

$$\Xi = \prod_k \sum_{\eta=0}^{\max N_k} e^{[-(\beta \varepsilon_k + \gamma)\eta]}, \tag{25.9}$$

in which the product is taken over all allowed single-particle quantum states and η is the number of particles in each state. (This process is similar to that used in the Debye treatment of the crystalline solid, in which we wrote

$$Q_v = \prod_v q_v = \prod_v \sum_n e^{-(n+\frac{1}{2})h\nu/kT}, \tag{25.10}$$

taking the product of the contributions from each of the allowed vibrational modes.)

This form for Ξ is convenient for use in dealing with systems in which we must take explicit account of quantum mechanical limitations on the allowed number of particles per state. For the case considered here, that of electrons which obey Fermi–Dirac statistics, with no more than one particle per quantum state, the only allowed values of η are 0 and 1. Thus the sum in equation (25.9) will contain only two terms, and we will have

$$\sum_{\eta=0}^{\max N_k} e^{-(\beta\varepsilon_k+\gamma)\eta} = 1 + e^{-\beta\varepsilon_k-\gamma}. \quad (25.11)$$

This leads to

$$\Xi = \prod_k (1 + e^{-\beta\varepsilon_k-\gamma}), \quad (25.12)$$

or

$$\ln \Xi = \sum_k \ln(1 + e^{-\beta\varepsilon_k-\gamma}). \quad (25.13)$$

This is a general expression that will apply to *any* system obeying Fermi–Dirac statistics.

AVERAGE NUMBER OF PARTICLES PER STATE

Before we apply the expression for Ξ to the case of the electron gas, it is instructive to use it to determine the average number of particles per quantum state, $\bar{N}_k$, in any system that follows Fermi–Dirac statistics.

In general,

$$\bar{N} = -\left(\frac{\partial \ln \Xi}{\partial \gamma}\right)_{\beta,V}. \quad (25.14)$$

For the present case, this leads to

$$\bar{N} = \sum_k \frac{1}{e^{\beta\varepsilon_k+\gamma} + 1} \quad (25.15)$$

for the system as a whole. The expected number of particles in a given single-particle state k, or, alternatively, the probability of occupancy of a given single particle state k, is thus

$$\bar{N}_k = \frac{1}{e^{\beta\varepsilon_k+\gamma} + 1}$$

$$= \frac{1}{\exp\left(\dfrac{\varepsilon_k - \mu}{kT}\right) + 1}. \quad (25.16)$$

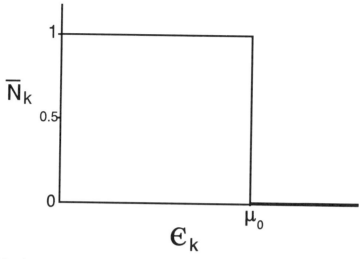

Figure 25.2 Average number of electrons per allowed energy state versus state energy, ε_k, at $T = 0$ K.

If we look at the limits of this expression as $T \to 0$, we see that the exponent in the denominator approaches infinity for $\varepsilon_k > \mu$, and that it approaches minus infinity for $\varepsilon_k < \mu$. [That is, $\exp(\infty) \to \infty$, $\exp(-\infty) \to 0$.] This leads to

$$\bar{N}_k = 1 \quad \text{for } \varepsilon_k < \mu,$$
$$\bar{N}_k = 0 \quad \text{for } \varepsilon_k > \mu. \qquad (25.17)$$

That is, at $T \to 0$, all states up to $\varepsilon_k = \mu$ are always occupied, all states above $\varepsilon_k = \mu$ are always empty, as shown in Figure 25.2. We will return to the form of this plot for $T > 0$ later.

EVALUATION OF THE THERMODYNAMIC FUNCTIONS AT 0 K

Let us return now to the evaluation of the expression for $\ln \Xi$ to determine the thermodynamic functions for the electron gas. Evaluation of the sum shown in equation (25.13) is difficult to carry out for the general case. However, we may determine much useful information about the system by looking at the behavior of $\ln \Xi$ and its derivatives with respect to μ, V, and T in the limit where $T \to 0$. This is a consequence of the fact, noted previously, that the kinetic energies of most of the free electrons will be large compared to kT for any temperature at which the metal will be a solid. As a result, the contributions of the free electrons to the thermodynamic properties will not differ greatly from the values at 0 K for any temperature of practical interest.

EVALUATION OF THE THERMODYNAMIC FUNCTIONS AT 0 K 323

As we have shown above, as $T \to 0$, each term in the expression for $\ln \Xi$ either vanishes or approaches infinity, dependent on the sign of $(\beta \varepsilon_k + \gamma)$.

To work with this, we will define

$$E_0 = -\lim_{T \to 0} \left(\frac{\partial \ln \Xi}{\partial \beta}\right)_{\gamma, V} = \lim_{T \to 0} \sum_k \bar{N}_k \varepsilon_k, \qquad (25.18)$$

$$N_0 = -\lim_{T \to 0} \left(\frac{\partial \ln \Xi}{\partial \gamma}\right)_{\beta, V} = \lim_{T \to 0} \sum_k \bar{N}_k, \qquad (25.19)$$

$$P_0 = \lim_{T \to 0} \left(\frac{1}{\beta}\right)\left(\frac{\partial \ln \Xi}{\partial V}\right)_{\beta, \gamma} = -\lim_{T \to 0} \sum_k \bar{N}_k \left(\frac{d\varepsilon_k}{dV}\right). \qquad (25.20)$$

In these expressions, as before,

$$\bar{N}_k = \frac{1}{e^{\beta \varepsilon_k + \gamma} + 1} = \frac{1}{e^{(\varepsilon_k - \mu)/kT} + 1}. \qquad (25.21)$$

At this point it is convenient to make a change of variables by defining

$$\varepsilon'_k = \varepsilon_k - u_0 = \left(\frac{h^2}{8mV^{2/3}}\right)(k_1^2 + k_2^2 + k_3^2), \qquad (25.22)$$

which is simply the translational part of ε_k. The variation of ε'_k with V, which is the derivative that we need in order to evaluate P_0, is

$$\frac{d\varepsilon'_k}{dV} = \left(\frac{h^2}{8m}\right)(k_1^2 + k_2^2 + k_3^2) \frac{d}{dV}\left(\frac{1}{V^{2/3}}\right)$$

$$= \left(-\frac{2}{3}\right)\left(\frac{h^2}{8mV^{2/3}}\right)(k_1^2 + k_2^2 + k_3^2)\left(\frac{1}{V}\right)$$

$$= -\frac{2}{3}\left(\frac{\varepsilon'_k}{V}\right). \qquad (25.23)$$

This leads to

$$P_0 = -\lim_{T \to 0} \sum_k \bar{N}_k \left(\frac{d\varepsilon'_k}{dV}\right)$$

$$= \frac{2}{3}\left(\frac{1}{V}\right) \lim_{T \to 0} \sum_k \bar{N}_k \varepsilon'_k, \qquad (25.24)$$

or

$$P_0 V = \frac{2}{3} \lim_{T \to 0} \sum_k \bar{N}_k \varepsilon'_k, \qquad (25.25)$$

We may alternatively write this expression in terms of a sum over energies rather than a sum over k-values, using the concept of a quantum number space just as we did with the monatomic ideal gas, yielding

$$P_0 V = \frac{2}{3} \lim_{T \to 0} \sum_{\varepsilon'} g(\varepsilon') N_k \varepsilon'$$

$$N_0 = \lim_{T \to 0} \sum_{\varepsilon'} g(\varepsilon') N_k. \tag{25.26}$$

In these expressions $g(\varepsilon')$, except for the factor of two arising from the degeneracy of electron quantum states, is the same as that developed for the monatomic ideal gas. That is,

$$g(\varepsilon') = \frac{\pi}{2} \left(\frac{8m}{h^2}\right)^{3/2} V \varepsilon'$$

$$= 4\pi \left(\frac{2m}{h^2}\right)^{3/2} V \varepsilon'. \tag{25.27}$$

This yields

$$P_0 V = \frac{2}{3} \lim_{T \to 0} \sum_{\varepsilon'} 4\pi \left(\frac{2m}{h^2}\right)^{3/2} V \varepsilon'^{3/2} N_k,$$

$$N_0 = \lim_{T \to 0} \sum_{\varepsilon'} 4\pi \left(\frac{2m}{h^2}\right)^{3/2} \varepsilon'^{1/2} N_k. \tag{25.28}$$

In these two expressions, the only temperature-dependent terms are the N_k. As we are only interested, at the moment, in the behavior at $T \to 0$, it is convenient to define

$$\mu_0' = \mu_0 - u_0. \tag{25.29}$$

Using this definition, we may say that $N_k \to 1$ for $\varepsilon' < \mu_0'$ and $N_k \to 0$ for $\varepsilon' > \mu_0'$. This permits us to replace the sums in equation (25.28) by integrals, because the difference in successive terms will be small compared to the range of integration, and to set the upper limit of integration at μ_0'. Performing the required integrations yields

$$P_0 V = \left(\frac{8\pi}{3}\right)\left(\frac{2m}{h^2}\right)^{3/2} V \int_0^{\mu_0'} \varepsilon'^{3/2} d\varepsilon'$$

$$= \left(\frac{16\pi}{15}\right)\left(\frac{2m}{h^2}\right)^{3/2} V \mu_0'^{5/2}. \tag{25.30}$$

and

$$N_0 = (4\pi) \left(\frac{2m}{h^2}\right)^{3/2} V \int_0^{\mu_0'} \varepsilon'^{1/2} d\varepsilon'$$

$$= \left(\frac{8\pi}{3}\right) \left(\frac{2m}{h^2}\right)^{3/2} V\mu_0'^{3/2}. \tag{25.31}$$

We may evaluate E_0 by using

$$E_0 = \lim_{T \to 0} \sum_k N_k \varepsilon_k'$$

$$= \lim_{T \to 0} \left(\sum_k N_0 u_0 + \sum_k N_k \varepsilon_k'\right)$$

$$= N_0 u_0 + \tfrac{3}{2} P_0 V. \tag{25.32}$$

Finally, since at 0 K only one *system* quantum state is available (N_0 electrons are in the N_0 states of lowest ε, one electron per state) we have

$$S_0 = 0. \tag{25.33}$$

Consequently

$$F_0 = E_0 - TS_0 = E_0. \tag{25.34}$$

This set of expressions for the thermodynamic functions, while correct, is not the most useful. We may write expressions for all of the thermodynamic functions in terms of the more convenient independent variables N, V, and T by first evaluating μ_0' in terms of N and V, using equations (25.30) and (25.31), and then using the classical relations among the various functions. Comparing equations (25.30) and (25.31), we have

$$\mu_0' = \frac{5}{2} \frac{P_0 V}{N_0}, \tag{25.35}$$

or, substituting for $P_0 V$ from equation (25.30) and rearranging terms,

$$\mu_0' = \left(\frac{h^2}{2m}\right) \left(\frac{3}{8\pi}\right)^{3/2} \left(\frac{N}{V}\right)^{3/2}. \tag{25.36}$$

This value may be substituted into the previously-derived expressions for the thermodynamic functions to yield

$$P_0 V = \frac{1}{20} \left(\frac{h^2 N}{m}\right) \left(\frac{3N}{\pi V}\right)^{2/3}, \tag{25.37}$$

$$E_0 = N u_0 + \frac{3}{40} \left(\frac{h^2 N}{m}\right) \left(\frac{3N}{\pi V}\right)^{2/3}, \tag{25.38}$$

and, as before

$$S_0 = 0, \tag{25.39}$$

$$F_0 = E_0 = N u_0 + \frac{3}{40} \left(\frac{h^2 N}{m}\right) \left(\frac{3N}{\pi V}\right)^{2/3}, \tag{25.40}$$

and finally

$$\mu_0 = \frac{(F_0 - TS_0)}{N}$$

$$= u_0 + \left(\frac{h^2}{2m}\right)\left(\frac{3}{8\pi}\right)^{2/3}\left(\frac{N}{V}\right)^{2/3}$$

$$= u_0 + \mu'_0, \tag{25.41}$$

in agreement with equation (25.29).

THE FERMI ENERGY AND THE WORK FUNCTION

The energy μ'_0, which is the kinetic energy of the most energetic electron in the crystal at 0 K, is used often in the electron theory of metals and, as mentioned earlier, is called the *Fermi energy* and is given by

$$\mu'_0 = \left(\frac{h^2}{2m}\right)\left(\frac{3N}{8\pi V}\right)^{2/3}. \tag{25.42}$$

Note that it is a function of N/V only. Typical values of μ'_0 are of the order of several electron volts. Another much-used parameter is the *work function*, Φ, which is defined as the work required to remove an electron from the metal at 0 K. This is related to the Fermi energy by

$$\Phi = -\mu_0 = -\mu'_0 - u_0. \tag{25.43}$$

Note that since electrons do not spontaneously escape from a metal at 0 K, μ_0 must be negative, and μ'_0 must always be smaller than u_0.

TEMPERATURE DEPENDENCE OF THE THERMODYNAMIC FUNCTIONS

Let us turn now to the question of describing the thermodynamic behavior of the free electron gas at tempertures above 0 K. If we look at the number of electrons in a given energy range, we see that the *maximum allowed* number of electrons at any energy ε'_k above u_0 is set by the number of available states at that value of ε'_k—that is, by $g(\varepsilon')$. The total number of electrons *actually present* in any range at a given temperature will be less than or equal to $g(\varepsilon')$ for that range, depending on the value of $\bar{N}_k$ for that particular value of ε'_k. At this point, we may rewrite the expression for $\bar{N}_k$ given in equation (25.16) in terms of the parameters ε'_k and μ'_0 as

$$\bar{N}_k = \frac{1}{[e^{(\varepsilon'_k - \mu'_0)/kT} + 1]}. \tag{25.44}$$

A plot of this equation is shown in Figure 25.3 for the case where $T = 800$ K and $\mu'_0 = 7.04$ eV, the value appropriate to copper. We see that this figure differs from Figure 25.2 in that we now have some states just below μ'_0 that have $\bar{N}_k < 1$, and some states just above μ'_0 that have $\bar{N}_k > 0$. If we now multiply $\bar{N}_k$ by $g(\varepsilon')$ and normalize by dividing by N, the total number of free electrons in the system, we can see that the fraction of the electrons in any given energy range is

$$\frac{1}{N}\left(\frac{dN}{d\varepsilon}\right) = \frac{\bar{N}_k g(\varepsilon'_k)}{N}. \tag{25.45}$$

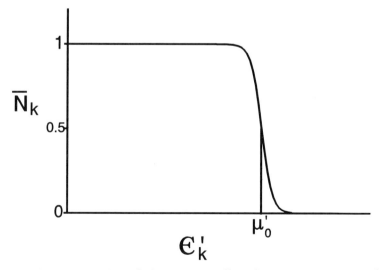

Figure 25.3 Average number of electrons per allowed energy state versus electron kinetic energy, ε'_k, for $T = 800$ K and $\mu'_0 = 7.04$ eV.

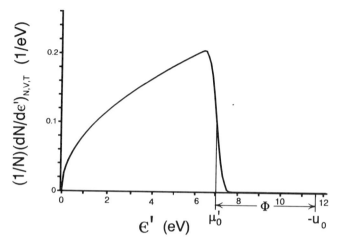

Figure 25.4 Fraction of the free electrons in a given kinetic energy range as a function of electron kinetic energy, ε'_k, for $T = 800$ K and $\mu'_0 = 7.04$ eV.

By substituting for $\bar{N}_k$ and $g(\varepsilon')$ from equations (25.44) and (25.27), we obtain

$$\left[\frac{1}{N}\left(\frac{dN}{d\varepsilon'}\right)\right]_{N,V,T} = \frac{4\pi\left(\frac{2m}{h^2}\right)^{3/2}(\varepsilon')^{1/2}V}{(e^{(\varepsilon'-\mu_0)/kT}+1)N}. \qquad (25.46)$$

The parameter on the left-hand side of this equation is the normalized density of states for the system. This function is plotted for copper at 800 K in Figure 25.4. It can be seen from the figure that because $\bar{N}_k$ has finite values over a range of values of ε'_k around μ'_0, the cutoff between filled and empty states is not sharp, leading to the shape of the curve shown in the figure. Consequently, the values of the various thermodynamic functions at finite temperatures will also differ from the values at 0 K.

We will not go through a rigorous treatment of the form of the expressions for the various thermodynamic functions at finite temperature, but will merely state approximate results in terms of a series expansion based on μ'_0, the Fermi energy. The appropriate expressions are

$$F' = F'_0\left[1 - \frac{5\pi^2}{12}\left(\frac{kT}{\mu'_0}\right)^2 + \cdots\right], \qquad (25.47)$$

$$E' = E'_0\left[1 + \frac{5\pi^2}{12}\left(\frac{kT}{\mu'_0}\right)^2 + \cdots\right], \qquad (25.48)$$

$$\mu' = \mu'_0 \left[1 - \frac{\pi^2}{12} \left(\frac{kT}{\mu'_0} \right)^2 + \cdots \right], \tag{25.49}$$

$$S = \frac{\pi^2}{2} Nk \left(\frac{kT}{\mu'_0} \right) + \cdots, \tag{25.50}$$

$$P = P_0 \left[1 + \frac{5\pi^2}{12} \left(\frac{kT}{\mu'_0} \right)^2 + \cdots \right], \tag{25.51}$$

$$C_V = \frac{\pi^2}{2} Nk \left(\frac{kT}{\mu'_0} \right) + \cdots. \tag{25.52}$$

In these expressions

$$F'_0 = E'_0 = E_0 - Nu_0. \tag{25.53}$$

For most metals, at temperatures below the melting point, $kT \ll \mu'_0$. Consequently the temperature-dependent terms in the above equations are small compared to unity. Thus the contributions of the free electrons to system properties such as S and C_V are small. This result is one of the major accomplishments of the thermodynamic treatment, because it shows why C_V for the electron gas does not have the classical value of $\frac{3}{2} R$ per mole found for the ideal monatomic gas. The behavior of S, C_V, and E as a function of temperature are shown in Figure 25.5, again for the case of copper. Note that the departures from the 0 K values are in all cases small at temperatures below the melting point, T_m.

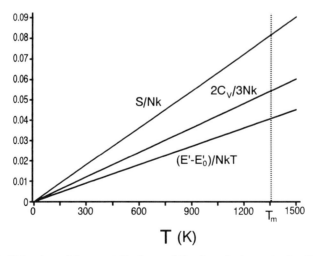

Figure 25.5 Behavior of the contributions of the free electrons to the thermodynamic properties of a solid, as a function of temperature, for the case of copper, for which $\mu'_0 = 7.04\,\text{eV}$.

BIBLIOGRAPHY

D. L. Goodstein, *States of Matter*, Prentice-Hall, Englewood Cliffs, NJ, 1975, Chapter 2.

T. L. Hill, *Introduction to Statistical Thermodynamics*, Addison-Wesley, 1960, Chapter 10.

C. C. Kittel and H. Kroemer, *Thermal Physics*, 2nd ed., W. H. Freeman & Co., San Francisco, 1980, Chapter 9.

E. L. Knuth, *Introduction to Statistical Thermodynamics*, McGraw-Hill, New York, 1966, Chapter 11.

W. G. V. Rosser, *An Introduction to Statistical Physics*, Ellis Horwood, Chichester, UK, 1982, Chapters 11 and 12.

PROBLEMS

25.1 The contribution to the heat capacity of a metal by the free electrons has been found to be

$$\left(\frac{C_V}{T}\right)_{electronic} = 5 \times 10^{-4} \, \text{J/mol-K}^2.$$

Calculate the Fermi energy and the effective number of free electrons per atom, assuming that the molar volume of the metal is $10 \, \text{cm}^3/\text{mol}$.

25.2 In what temperature range would it be appropriate to use the Boltzmann approximation to calculate the properties of the electron gas in a typical metal?

25.3 Calculate, for the free electrons in silver, the values of the Fermi energy, μ'_0, the pressure, P_0, and the energy E_0/N_0.

25.4 Calculate the velocity of an electron in copper that has a value of $\varepsilon' = \mu'_0$.

CHAPTER 26

BLACKBODY RADIATION

The final system that we will consider is that of blackbody radiation, or, as it is sometimes called, the photon gas. In this case, we will again be treating a gas of noninteracting particles (or, alternatively, a system of standing waves) in an enclosure, but with the complication, similar to that encountered in Chapter 25, that the conditions for the use of Boltzmann statistics are not met. The present case differs from that of Chapter 25, however, in that photons have unity spin, and therefore follow Bose–Einstein statistics.

THE MODEL

The model that we will use in this case consists of an enclosure of volume $V = l^3$, with walls that are perfect reflectors for electromagnetic radiation (photons), except for a small black area that is a perfect absorber and reradiator of photons, and is maintained at a constant temperature by contact with an external reservoir. This black area serves to maintain thermal equilibrium between the system and the reservoir, as well as among the electromagnetic waves (photons) within the enclosure.

Note that the total number of photons in the system is *not* conserved, because the small black area serves essentially as a "catalyst" for the reaction

$$nv_i \rightleftharpoons mv_j. \tag{26.1}$$

in which a number, n, of photons of energy hv_i are absorbed and some other number, m, photons of energy hv_j are emitted, subject to the constraint that the total energy of the ensemble plus reservoir is unchanged. As a consequence of this, there is no constraint on the total number of photons in an ensemble made up of systems like this, and as a result the parameter γ that arose when we considered the interchange of particles between system and surroundings in the development of the grand canonical ensemble partition function, Ξ, has no meaning in this case. In that case, in Chapter 7, we were implicitly considering a process in which the number of particles in one system of the ensemble increased by an amount dN_i and the number of particles in another system of the ensemble decreased by that same amount, so that the total number of particles in the ensemble plus reservoir was conserved. It was this constraint on the total number of particles that led to the parameter γ. In the present case, the lack of a constraint on the total number of particles means that γ is undetermined, and that it, and the classical equivalent parameter μ, can be set equal to zero for this system.

EVALUATION OF THE PARTITION FUNCTION

As we have done in previous cases, we may determine whether or not the conditions required for the application of Boltzmann statistics are met in this system by comparing the number of available single-particle quantum states with the expected number of particles in the system, to assess the likelihood of finding two particles in the same quantum state. The number of available states may be determined, as in previous cases that we have considered, using the concept of a three-dimensional quantum number space to determine the number of states having energies less than or equal to kT. If we look at this system in terms of the wave description of the photon—that is, as a system of standing waves in the enclosure—the problem of determining the allowed wavelengths in this case is identical to that considered for the Debye model of the solid. This calculation, in Chapter 15, led to the expression

$$G(\lambda) = \frac{4\pi}{3}\left(\frac{l^3}{\lambda^3}\right). \tag{26.2}$$

To go to an expression in terms of the number of allowed states up to a given frequency, we must use the relation

$$v = \frac{c}{\lambda}, \tag{26.3}$$

in which c is the velocity of light, and must take account of the fact that, for each frequency, there are two allowed states, of different polarizations. This leads to the final relation

$$G(v) = \frac{8\pi}{3}\left(\frac{Vv^3}{c^3}\right). \tag{26.4}$$

EVALUATION OF THE PARTITION FUNCTION 333

The number of states having an energy $h\nu \leq kT$ is thus

$$G\left(\frac{kT}{h}\right) = \frac{8\pi V k^3 T^3}{3h^3 c^3}. \tag{26.5}$$

This must be compared to the expected number of photons in a system of volume V at temperature T. We will see later in this chapter that

$$\bar{N} = \frac{16\zeta(3)V k^3 T^3}{h^3 c^3}, \tag{26.6}$$

where the zeta function of three, $\zeta(3)$, equals 1.202. Thus

$$\frac{G\left(\frac{kT}{h}\right)}{\bar{N}} = \frac{\frac{8\pi V k^3 T^3}{3h^3 c^3}}{\frac{16\zeta(3)V k^3 T^3}{h^3 c^3}} = \frac{8\pi}{48\zeta(3)} < 1, \tag{26.7}$$

and the conditions for use of Boltzmann statistics are clearly not met.

Because we must take explicit account of the number of photons per quantum state, we will again use the partition function Ξ, as we did in Chapter 25. In this case, as before, we may write the general expression

$$\Xi = \prod_k \sum_{\eta=0}^{\max N_k} e^{-(\beta\varepsilon_k + \gamma)\eta}. \tag{26.8}$$

However, in this case, the value of max N_k will be infinity, rather than one, and we have already determined that $\gamma = 0$ for this system. The sum in the expression for Ξ is thus

$$\sum_{\eta=0}^{\max N_k} e^{-(\beta\varepsilon_k + \gamma)\eta} = \sum_{\eta=0}^{\infty} e^{(-\beta\varepsilon_k)\eta} = \sum_{\eta=0}^{\infty} x^\eta, \tag{26.9}$$

where in the final step we have defined

$$x \equiv e^{-\beta\varepsilon_k}. \tag{26.10}$$

It can be shown on a purely mathematical basis that this series converges if $x < 1$. In the present case, since β and ε_k are both finite and positive, this condition is met, and we have

$$\sum_{\eta=0}^{\infty} x^\eta = \frac{1}{1-x}. \tag{26.11}$$

Thus we have

$$\Xi = \prod_k (1 - e^{-\beta\varepsilon_k})^{-1},$$

$$\ln \Xi = \sum_k (1 - e^{-\beta\varepsilon_k})^{-1}. \tag{26.12}$$

We may begin the evaluation of this sum by restating it in terms of a sum over frequency increments, rather than a sum over k values, as

$$\ln \Xi = \sum_\nu g(\nu) \ln \frac{1}{1 - e^{-\beta h\nu}}, \tag{26.13}$$

where

$$g(\nu) = \frac{dG(\nu)}{d\nu} = \frac{d}{d\nu}\left(\frac{8\pi V}{3c^3}\nu^3\right) = \frac{8\pi V}{c^3}\nu^2. \tag{26.14}$$

The evaluation of this sum will be simplified if we can replace the sum with an integral, as we have done previously. Again, this will be justified if the difference in the values of successive terms in the sum is small compared to unity. This will be so in the present case if

$$\beta h \Delta \nu = \frac{\beta h}{2lkT} \tag{26.15}$$

is small compared to unity. For the case of $T = 1$ K and $l = 1$ cm, we determine that $hc/2lkT \approx 1$, so that at any temperature of practical interest the condition for use of an integral is met, and we may write

$$\ln \Xi = \int_0^\infty g(\nu) \ln\left(\frac{1}{1 - e^{-\beta h\nu}}\right) d\nu$$

$$= \frac{8\pi V}{c^3} \int_0^\infty \ln\left(\frac{1}{1 - e^{-\beta h\nu}}\right) \nu^2 \, d\nu. \tag{26.16}$$

This may be integrated by parts to yield

$$\ln \Xi = \frac{8\pi V \nu^3}{3c^3} \ln \frac{1}{1 - e^{-\beta h\nu}} \bigg|_0^\infty + \frac{8\pi V \beta h}{3c^3} \int_0^\infty \frac{\nu^3}{e^{\beta h\nu} - 1} d\nu. \tag{26.17}$$

The first term on the right is zero, as its value is zero at each limit. We may evaluate the second term on the right by introducing the dummy variables

$$x \equiv \beta h \nu, \qquad dx \equiv \beta h \, d\nu, \qquad (26.18)$$

to yield

$$\ln \Xi = \frac{8\pi V}{3c^3(\beta h)^3} \int_0^\infty \frac{x^3}{e^x - 1} dx. \qquad (26.19)$$

The value of this definite integral is $6\zeta(4) = \pi^4/15$, leading to the final expression for Ξ of

$$\ln \Xi = \frac{8\pi V}{3(\beta h c)^3} \frac{\pi^4}{15} = \frac{8\pi^5 V}{45(\beta h c)^3}, \qquad (26.20)$$

or

$$\ln \Xi = \frac{8\pi^5 V k^3 T^3}{45 h^3 c^3}. \qquad (26.21)$$

Note that $\Xi = \Xi(V, T)$ only, since we have shown that $\mu = 0$ for this system.

EVALUATION OF THE THERMODYNAMIC FUNCTIONS

From this point we may proceed as usual to evaluate the thermodynamic functions of the photon gas in terms of $\ln \Xi$ and its partial derivatives. Evaluation of most of the state functions is straightforward, because we may use the relations between $\ln \Xi$ and the thermodynamic functions developed in Chapter 9. Application of these relations leads to

$$P = kT \left(\frac{\partial \ln \Xi}{\partial V} \right)_T$$

$$= \frac{8\pi^5 k^4 T^4}{45 h^3 c^3} \frac{\partial}{\partial V}(V)$$

$$= \frac{8\pi^5 k^4 T^4}{45 h^3 c^3}. \qquad (26.22)$$

$$E = kT^2 \left(\frac{\partial \ln \Xi}{\partial T} \right)_V$$

$$= \frac{8\pi^5 V k^4}{45 h^3 c^3} \frac{\partial}{\partial T}(T^5)$$

$$= \frac{8\pi^5 V k^4 T^4}{15 h^3 c^3}. \qquad (26.23)$$

336 BLACKBODY RADIATION

$$C_V = \left(\frac{\partial E}{\partial T}\right)_V$$

$$= \frac{\partial}{\partial T}\left(\frac{8\pi V k^4 T^4}{15 h^3 c^3}\right)$$

$$= \frac{35\pi^5 V k^4 T^3}{15 h^3 c^3}. \tag{26.24}$$

$$S = kT\left(\frac{\partial \ln \Xi}{\partial T}\right) + k \ln \Xi = \frac{E}{T} + \frac{PV}{T}$$

$$= \frac{1}{T}\left(\frac{8\pi^5 V k^4 T^4}{15 h^3 c^3}\right) + \frac{V}{T}\left(\frac{8\pi^5 k^4 T^4}{45 h^3 c^3}\right)$$

$$= \frac{32\pi^5 V k^4 T^3}{45 h^3 c^3}. \tag{26.25}$$

Note that the power dependences of E, S, and C_V on temperature are the same as those deduced in Chapter 15 for the Debye model of the solid at low temperatures. This arises from the fact that both systems may be treated as a system of standing waves in an enclosure, where in the case of the Debye crystal the enclosure is the crystal itself. The major differences are that, in the case of the crystal, we have three modes for each wavelength, two transverse and one longitudinal, rather than the two transverse modes in the photon gas, and that, in the crystal, all modes may have different velocities of propagation, while all photons propagate at the speed of light. Because of this similarity, the Debye model is sometimes referred to as a *phonon gas*—that is, a gas composed of acoustical quanta, rather than the optical quanta of the photon gas.

Returning to the problem of determining the values of the thermodynamic functions, we see that a problem arises when we try to calculate a value for the expected number of photons in the system, because the general relation between N and $\ln \Xi$ developed in Chapter 9 is

$$N = kT\left(\frac{\partial \ln \Xi}{\partial \mu}\right)_{V,T}, \tag{26.26}$$

and we have set $\mu = 0$. We may circumvent this problem by first returning to the general expression for $\ln \Xi$ for systems following Bose–Einstein statistics, namely

$$\ln \Xi = \sum_k \ln(1 - e^{-\beta \varepsilon_k + \gamma})^{-1}, \tag{26.27}$$

differentiating it with respect to γ to yield

$$N = -\left(\frac{\partial \ln \Xi}{\partial \gamma}\right) = \sum_k (e^{\beta \varepsilon_k + \gamma} - 1)^{-1}, \qquad (26.28)$$

then setting $\gamma = 0$ and recasting the expression for N from a sum over quantum states to an integral over frequencies to yield

$$N = \sum_k (e^{\beta \varepsilon_k} - 1) = \sum_v g(v)(e^{\beta h v} - 1)^{-1}$$

$$= \int_0^\infty g(v)(e^{\beta h v} - 1)^{-1} dv = \frac{8\pi V}{c^3} \int_0^\infty \frac{v^2}{e^{\beta h v} - 1} dv, \qquad (26.29)$$

and then evaluating the integral using the dummy variables defined in equation (26.18) to yield, finally,

$$N = \frac{8\pi V k^3 T^3}{h^3 c^3} \int_0^\infty \frac{x^2}{(e^x - 1)} dx$$

$$= \frac{16\pi \zeta(3) V k^3 T^3}{h^3 c^3}, \qquad (26.30)$$

where $\zeta(3) = 1.202$ as before. Note that N increases rapidly and without limit as T increases. Because of this the correspondence between this system and the Debye treatment breaks down at higher temperatures: The number of allowed vibrational modes in the crystal is limited to three times the number of atoms present, and thus the increases in the thermodynamic properties with temperature must saturate once all of these modes are excited.

THE SPECTRAL ENERGY DISTRIBUTION

We will make one final manipulation with the thermodynamic properties of the photon gas to demonstrate the practical consequences of the treatment. We will determine the distribution of the energy of the system among the frequencies present, or, as it is sometimes called, the spectral energy distribution. This distribution is important both in the practical context of radiant heat transfer and in terms of the relation of the quantum mechanical treatment presented here to earlier classical treatments of the same problem.

To determine the spectral energy distribution, which is essentially dE/dv, we begin with the expression developed in equation (26.29),

$$N = \int_0^\infty g(v)(e^{\beta h v} - 1)^{-1} dv = \frac{8\pi V k^3 T^3}{h^3 c^3} \int_0^\infty \frac{\left(\frac{hv}{kT}\right)^2}{e^{hv/kT} - 1} d\left(\frac{hv}{kT}\right), \qquad (26.31)$$

and differentiate it with respect to hv/kT at constant T and V to yield

$$\left(\frac{\partial N}{\partial \left(\frac{hv}{kT}\right)}\right)_{V,T} = \frac{8\pi V k^3 T^3}{h^3 c^3} \frac{\left(\frac{hv}{kT}\right)^2}{(e^{hv/kT} - 1)}. \quad (26.32)$$

The relationship between a change in N and the corresponding change in E when a photon of frequency v is added to or removed from the system is

$$\left(\frac{\partial \left(\frac{E}{kT}\right)}{\partial N}\right)_{V,T} = \frac{hv}{kT}. \quad (26.33)$$

Multiplication of equation (26.32) by equation (26.33) yields

$$\left(\frac{\partial \left(\frac{E}{kT}\right)}{\partial \left(\frac{hv}{kT}\right)}\right)_{V,T} = 8\pi V \left(\frac{kT}{hc}\right)^3 \frac{\left(\frac{hv}{kT}\right)^3}{(e^{hv/kT} - 1)}. \quad (26.34)$$

This expression is known as *Planck's formula* and is shown in Figure 26.1, plotted in dimensionless form. Note that photons of low frequency make a

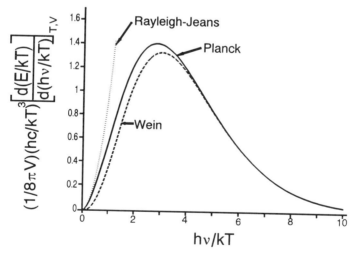

Figure 26.1 Plot of the dimensionless spectral energy associated with a photon gas, as a function of the dimensionless frequency, hv/kT. The solid line describes the behavior according to Planck's formula, equation (26.34). The dotted line represents the classical Rayleigh–Jeans curve; the dashed line represents the Wien formula.

relatively small contribution total energy, due to the small energy per photon, while high-frequency photons also make a small contribution due to their relatively small numbers. The maximum contribution comes from photons in the frequency range about the maximum in the curve, which occurs at $h\nu/kT = 2.822$. The total area under the curve is simply the total energy of the system, which may be verified by integrating the energy distribution over $h\nu/kT$ values from zero to infinity.

If we look at the effect of temperature on the properties of the system, we see that, since the total energy is proportional to T^4, doubling the temperature increases the total energy by a factor of 16. The frequency associated with the maximum in the curve will increase by a factor of 2, and the height of the maximum will increase by a factor of 8 for this same temperature change. This increase in the frequency of maximum intensity is known as *Wien's displacement law*.

Finally, let us look at the high and low temperature limits of the Planck formula. At low temperature, the exponential in the denominator of equation (26.34) will be large compared to unity, leading to

$$\left(\frac{\partial \left(\frac{E}{kT} \right)}{\partial \left(\frac{h\nu}{kT} \right)} \right)_{V,T} \approx 8\pi V \left(\frac{kT}{hc} \right)^3 \frac{\left(\frac{h\nu}{kT} \right)^3}{e^{h\nu/kT}}. \tag{26.35}$$

This relation, which is also plotted in Figure 26.1, is known as the *Wien formula* and gives a reasonable approximation to the Planck formula at high frequencies, but is deficient at low frequencies. An equivalent formula was developed by Wien classically prior to the development of quantum mechanics.

At high temperatures, the limiting form of equation (26.34) is

$$\left(\frac{\partial \left(\frac{E}{kT} \right)}{\partial \left(\frac{h\nu}{kT} \right)} \right)_{V,T} \approx 8\pi V \left(\frac{kT}{hc} \right)^3 \left(\frac{h\nu}{kT} \right)^2. \tag{26.36}$$

This equation predicts that the energy density will increase without limit as $h\nu/kT$ increases. An equivalent formula was first developed by Rayleigh on a classical basis and is known as the *Rayleigh–Jeans formula*. It is also plotted in Figure 26.1. Note that the Rayleigh–Jeans formula, which was derived on the assumption that each photon mode had an energy kT, grossly overestimates the contribution of high-frequency modes to the system energy. The development of the Planck formula resolved this discrepancy between theory and observation and was also the first occasion in which the constant h appeared in a thermodynamic formula.

BIBLIOGRAPHY

E. A. Guggenheim, *Thermodynamics*, North Holland, 1957, Chapter 13.

T. L. Hill, *Introduction to Statistical Thermodynamics*, Addison-Wesley, 1960, Chapter 22.

C. C. Kittel and H. Kroemer, *Thermal Physics*, 2nd ed., W. H. Freeman & Co., San Francisco, 1980, Chapter 4.

E. L. Knuth, *Introduction to Statistical Thermodynamics*, McGraw-Hill, New York, 1966, Chapter 10.

W. G. V. Rosser, *An Introduction to Statistical Physics*, Ellis Horwood, Chichester, UK, 1982, Chapters 9, 11, and 12.

PROBLEMS

26.1 Verify that the area under the Planck curve is equal to the system energy given in equation (26.23).

26.2 Calculate the pressure exerted by blackbody radiation at temperatures of 300 K and 10^6 K (typical of a thermonuclear reaction), and compare the results to the pressure exerted by a monatomic ideal gas at the same temperatures and a density of 0.05 mol/liter.

26.3 At what temperature does the maximum in the Planck curve correspond to the midpoint of the visual range, which extends from roughly 390 nm to 780 nm?

26.4 At what temperature would the energy per unit volume in a radiation field equal the energy of one mole of a monatomic ideal gas at 300 K?

APPENDIX A

FUNDAMENTAL CONSTANTS AND CONVERSIONS

Fundamental Constants

Avogadro's number	N_{Av}	6.022×10^{23}
Molar gas constant	R	8.314 J/mol – K
Boltzmann's constant	k	1.381×10^{-23} J/K
Atomic mass unit	u	1.661×10^{-27} kg
Planck's constant	h	6.626×10^{-34} J-sec
$h/2\pi$	$\hbar$	1.055×12^{-34} J-sec
Elementary charge	e	1.602×10^{-19} coul
Electron volt	eV	1.602×10^{-19} J
Permittivity of vacuum	ε_0	8.854 F/m
Speed of light	c	2.998×10^8 m/sec
Stefan–Boltzmann constant	σ	5.671×10^{-8} W/m²K⁴

Energy Conversion Factors

$1 \text{ J} = 0.2388 \text{ cal} = 1 \times 10^7 \text{ erg} = 9.869 \times 10^{-3} \text{ liter-atm} = 6.242 \times 10^{18} \text{ eV}.$

The Gas Constant

$R = 8.314 \text{ J/mol-K} = 8.314 \text{ Pa-m}^3/\text{mol-K} = 0.08205 \text{ liter-atm/mol-K}.$

Boltzmann's Constant

$k = 1.381 \times 10^{-23} \text{ J/K} = 8.618 \times 10^{-5} \text{ eV/K} = 1.381 \times 10^{-16} \text{ erg/K}.$

APPENDIX B

OTHER ENSEMBLES

As was mentioned in Chapter 6, the ensembles that we treated in detail are not the only possible ensembles, but are the ones that provide the correspondences needed to connect the classical and statistical treatments of thermodynamics. It is, of course, possible to construct ensembles based on any set of three of the thermodynamic state functions, which are sufficient to define the state of the system on a macroscopic basis. We considered one other such ensemble in Problem 8.1. An extended listing of the partition functions for eight possible choices of independent variables and their correspondence to classical parameters is given below:

$$
\begin{aligned}
N, V, E \qquad & \frac{S}{k} = \ln \Omega \\
N, V, T \qquad & -\frac{F}{kT} = \ln \sum_i e^{-\beta E_i} \\
\mu, V, E \qquad & \frac{H}{kT} = \ln \sum_i e^{-\gamma N_i} \\
\mu, V, T \qquad & \frac{PV}{kT} = \ln \sum_i e^{-\beta E_i} e^{-\gamma N_i} \\
N, P, E \qquad & -\frac{PV}{kT} + \frac{S}{k} = \ln \sum_i e^{-\pi V_i} \\
N, P, T \qquad & -\frac{G}{kT} = \ln \sum_i e^{-\beta E_i} e^{-\pi V_i}
\end{aligned}
\qquad (\text{B.1})
$$

μ, P, E	$\dfrac{E}{kT} = \ln \sum_i e^{-\gamma N_i} e^{-\pi V_i}$
μ, P, T	$0 = \ln \sum_i e^{-\beta E_i} e^{-\gamma N_i} e^{-\pi V_i}$

In the above, π is equivalent to the classical parameter P/kT.

APPENDIX C

THERMODYNAMIC DATA FOR ONE-COMPONENT SYSTEMS

Tabulated below are the experimentally measured values for the melting and normal boiling points, T_m and T_b respectively, the molar heats (ΔH_i) and entropies (ΔS_i) of fusion (f) and vaporization (v), and the molar volume change on melting, ΔV_f, of a range of substances. In all cases, temperatures are reported in K, enthalpy values in J/mol, entropies in J/mol-K, and volumes in cm^3/mol. Melting temperatures, enthalpies of fusion, and volume changes on melting are experimentally measured values. Enthalpies of vaporization are average values obtained from vapor pressure measurements. Entropies are calculated using $\Delta S_f = \Delta H_f/T_m$, $\Delta S_v = \Delta H_v/T_b$.

THERMODYNAMIC DATA FOR ONE-COMPONENT SYSTEMS

Substance	T_m	ΔH_f	ΔV_f	T_b	ΔH_v	ΔS_f	ΔS_v
Ag	1,234	11,094	1.27	2,470	257,400	8.99	104.2
Al	933	10,676	1.32	2,793	290,470	11.46	104.0
Al_2O_3	2,318	109,000		3,250		47.02	
Ar	83	1,214	3.53	87	6,526	14.63	75.0
Au	1,336	12,686	1.14	3,130	342,400	9.50	109.4
BeO	2,523	71,000		4,170	538,100	28.14	129
Bi	544	10,885	−0.57	1,837	179,100	20.01	97.5
Cd	594	6,113	1.00	1,038	99,400	10.29	95.8
Cu	1,357	13,027	0.85	2,830	305,600	9.60	108
Fe	1,803	14,905	0.86	3,130	360,800	8.29	109.0
Ge	1,232	37,450	−0.67	3,103	328,900	30.4	106
H_2O	273	6,012	−0.0906	373	40,656	22.02	108.9
Mg	923	8,796	1.32	1,363	127,700	9.53	93.7
MgO	2.915	77,500		3,870		26.59	
Na	371	17,585	1.10	1,156	98,000	7.11	84.8
Ne	24.5	324		27	1,860	13.22	69.0
Ni	1,725	17,156	0.83	3,180	375,200	9.94	118
Pb	600	4,806	1.66	2,013	178,500	8.01	88.4
Pt	2,043	19,695	1.20	4,400	470,800	9.64	107
Si	1,685	50,710	−0.81	3,540	385,800	30.1	109
SiO_2	1,743	14,200		2,500		8.14	
Sn	505	6,956	0.70	2,898	295,600	13.77	102
W	3,660	35,300	0.88	5,828	734,300	9.64	126
Xe	162	2,300	5.59	166	12,640	14.20	76.0

APPENDIX D

VAPOR PRESSURE RELATIONS

Empirical expressions for the temperature dependence of the equilibrium vapor pressure over both the solid and liquid phases of a wide variety of substances have been developed by a number of investigators. These expressions are of the form

$$\log_{10} P^0 = A + BT^{-1} + C \log_{10} T + DT^{-3}, \tag{C.1}$$

where T is in kelvins. The values of the constants A, B, C, and D are presented below for a variety of solid (s) and liquid (l) species. Use of these equations will yield the equilibrium vapor pressure in pascals.

Substance	A	B	C	D
Ag(s)	9.127	−14,999	−0.7845	
Ag(l)	5.752	−13,827		
Al(s)	9.459	−17,342	−0.7927	
Al(l)	5.911	−16,221		
Au(s)	9.152	−19,343	−0.7479	
Au(l)	5.832	−18,024		
BeO(s)	15.62	−34,230	−2	
Cd(s)	5.939	−5,799		
Cd(l)	5.242	−5,392		
Cu(s)	9.123	−17,748	−0.7317	
Cu(l)	5.894	−16,415		
Fe(s)	7,100	−21,723	0.4536	−0.5846
Fe(l)	6.347	−19,574		
Ge(s)	10.40	−20,150	−0.91	
Hg(l)	5.116	−3,190		
Mg(s)	8.489	−7,813	−0.8253	
Mg(l)	9.909	−7,550	−0.855	
Na(s)	5.298	−5,603		
Na(l)	4,704	−5,377		
NaCl(s)	11.43	−12,440	−0.90	
Ni(s)	10.557	−22,606	−0.8717	
Ni(l)	6.666	−20,765		
Pb(s)	5.643	−10,143		
Pb(l)	4.911	−9,701		
Pt(s)	4.882	−29,387	1.1039	−0.4527
Pt(l)	6.347	−26,856		
Si(s)	9.949	−18,000	−1.022	
Sn(s)	6.036	−15,710		
Sn(l)	5.262	−15,332		
Ti(s)	11.925	−24,991	−0.3142	
Ti(l)	6.358	−22,747		
W(s)	2.945	−44,094	1.3677	
Zn(s)	6.012	−6,776		
Zn(l)	5.378	−6,286		

More complete tabulations may be found in the *Handbook of Chemistry and Physics*, published by CRC Press, and in C. B. Alcock, V. P. Itkin, and M. K. Horrigan, *Canadian Metallurgical Quarterly*, **23**, 309 (1984).

APPENDIX E

MODEL POTENTIALS

There are many instances in which it is desired to treat the properties of a system in terms of the potential energy change associated with the interaction of one atom with another atom, or collection of atoms. In order to do this, it is necessary to have an analytic expression for the variation of system potential energy with the internuclear separation of a pair of atoms, or, as it is often referred to, a model potential. In this text, we have used model potentials for systems ranging from diatomic molecules to binary solid solutions.

A number of different model potentials have been used over a period of time. All of these have the common features of a repulsive component, which makes a positive contribution to system potential energy and which decreases very rapidly with increasing internuclear separation, and an attractive contribution, which makes a negative contribution to system potential energy and which decreases less rapidly with increasing internuclear separation. The exact form of these contributions differs from one model potential to another, reflecting, among other things, the nature of the attractive forces involved in the particular system being modeled.

One such potential is the Morse potential,

$$u(r) = \varepsilon[(1 - e^{-a(r-r_e)})^2 - 1], \tag{E.1}$$

developed by Morse as an empirical fit to spectroscopic data obtained for diatomic molecules. In this expression, ε is the well depth (D_e in the case of diatomic molecules considered by Morse), r_e is the equilibrium internuclear separation, and a is an empirical constant.

A more commonly used form for the potential assumes a power law dependence of both the attractive and repulsive parts of the potential, with the power dependence of the repulsive part being higher than that of the attractive

MODEL POTENTIALS   349

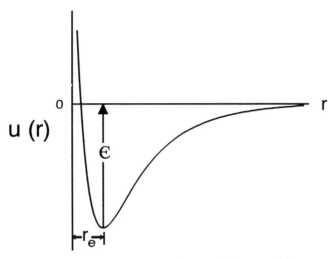

Figure E.1 The Lennard-Jones 6-12 potential.

part. A particularly common example of such a potential is that developed by Lennard-Jones, often referred to as the "6-12" potential, and given as

$$u(r) = -2\varepsilon \left(\frac{r_e}{r}\right)^6 + \varepsilon \left(\frac{r_e}{r}\right)^{12}, \tag{E.2}$$

where, again, ε, an inherently positive number, represents the maximum depth of the potential well, and r_e is the internuclear separation associated with this maximum depth. This potential is shown graphically in Figure E.1.

Because we make repeated use of the Lennard-Jones potential in this text, a summary of Lennard-Jones parameters determined for a variety of species is given below.

Substance	ε (J/mol)	r_e (Å)
He	84.97	2.869
H_2	307.6	3.287
Ne	290.1	3.12
N_2	790.2	4.151
O_2	981	3.88
CH_4	1232	4.825
Ar	996.0	3.40
Kr	1422	4.04
Xe	1873	4.07

For more information on intermolecular potentials, refer to J. O Hirschfelder, C. F. Curtiss, and R. B. Bird, *Molecular Theory of Gases and Liquids*, John Wiley & Sons, New York, 1954.

APPENDIX F

SPECTROSCOPIC DATA FOR DIATOMIC AND POLYATOMIC MOLECULES

Molecule	θ_v (K)	θ_r (K)	D_e (eV)	σ
H_2	6,210	85.4	4.54	2
D_2	4,300	42.7	4.54	2
N_2	3,340	2.86	7.37	2
O_2	2,230	2.07	5.08	2
CO	3,070	2.77	11.25	1
NO	2,690	2.42	5.29	1
HCl	4,140	15.2	4.43	1
HBr	3,700	12.1	3.60	1
HI	3,200	9.00	2.75	1
Cl_2	810	0.35	2.48	2
Br_2	470	0.12	1.97	2
I_2	310	0.05	1.54	2
H_2O	2,290 5,250 5,400	22.3	9.71	2
CO_2	960 (2) 2,000 3,380	0.56	16.77	2
N_2O	847 (2) 1,850 3,200	0.60	11.62	1
C_2H_2	880 (2) 1,050 (2) 2,840 4,730 4,850	1.69	17.10	2
NH_3	1,370 2,340 (2) 4,800 4,910 (2)	12.3	12.26	3

Note: In the case of polyatomic molecules, the number in parentheses indicates the number of vibrational modes having the same Θ_v. Values for D_e for polyatomic molecules represent the total potential energy associated with all of the interatomic bonds in the molecule and were calculated from thermodynamic data.

APPENDIX G

HYPERBOLIC FUNCTIONS

The following table of $\sinh x$ and $\coth x$ may be used in calculating the properties of crystalline solids using the Einstein model developed in Chapter 14, with $x = \Theta_E/2T$.

x	sinh x	coth x
0.00	0.0000	∞
0.10	0.1002	10.133
0.20	0.2013	5.0067
0.30	0.3045	3.4328
0.40	0.4108	2.6319
0.50	0.5211	2.1639
0.60	0.6367	1.8620
0.70	0.7586	1.6546
0.80	0.8881	1.5059
0.90	1.0265	1.3961
1.00	1.1752	1.3130
1.20	1.5095	1.1952
1.40	1.9043	1.1295
1.60	2.3756	1.0850
1.80	2.9422	1.0562
2.00	3.6269	1.0373
2.20	4.4571	1.0249
2.40	5.4662	1.0116
2.60	6.6947	1.0111
2.80	8.1919	1.0074
3.00	10.018	1.0050
3.50	16.543	1.0018
4.00	27.290	1.0007
4.50	45.003	1.0003
5.00	74.203	1.0001
5.50	122.34	1.0000
6.00	201.71	1.0000
6.50	332.57	1.0000
7.00	548.32	1.0000
7.50	904.02	1.0000
8.00	1490.5	1.0000
8.50	2457.4	1.0000
9.00	4051.5	1.0000
9.50	6679.9	1.0000
10.0	11013	1.0000

APPENDIX H

THE DEBYE FUNCTION AND THE DEBYE TEMPERATURE

In this appendix, we present values of the Debye function,

$$D(u) = \frac{3}{u^3} \int_0^u \frac{x^3}{e^x - 1} dx,$$

in which

$$u = \frac{\Theta_D}{T} = \frac{h\nu_{max}}{kT},$$

and measured values of Θ_D for a number of substances. This information may be used to calculate the thermodyamic properties of crystalline materials according to the Debye treatment presented in Chapter 15. Note that although the treatment in Chapter 15 implicitly considered crystals consisting of a single chemical species, the compilation of Θ_D values includes values determined for a number of compounds as well.

THE DEBYE FUNCTION

u	0.0	0.2	0.4	0.6	0.8
0.0	1.0000	0.9270	0.8580	0.7929	0.7318
1.0	0.6744	0.6208	0.5708	0.5243	0.4811
2.0	0.4411	0.4042	0.3701	0.3388	0.3100
3.0	0.2836	0.2594	0.2373	0.2170	0.1986
4.0	0.1817	0.1664	0.1524	0.1397	0.1281
5.0	0.1176	0.1080	0.09930	0.09137	0.08415
6.0	0.07758	0.07160	0.06615	0.06118	0.05664
7.0	0.05251	0.04873	0.04527	0.04211	0.03921
8.0	0.03656	0.03413	0.03189	0.02983	0.02794
9.0	0.02620	0.02459	0.02311	0.02174	0.02047
10.0	0.01930	0.01821	0.01720	0.01626	0.01538
11.0	0.01457	0.01381	0.01311	0.01245	0.01183
12.0	0.01125	0.01071	0.01020	0.00973	0.00928
13.0	0.00886	0.00846	0.00809	0.00774	0.00741
14.0	0.00710	0.00680	0.00652	0.00626	0.00601
15.0	0.00577	0.00555	0.00533	0.00513	0.00494

VALUES OF Θ_D

Substance	Θ_D (K)	Substance	Θ_D (K)
Ag	215	Mg	330
Al	398	Mo	375
Ar	85	Na	160
Au	185	Ne	63
B	1250	Ni	375
Be	980	Pb	86
C (diamond)	2230	Pt	225
Ca	230	Si	645
Cd	165	Ti	380
Cr	460	W	315
Cu	315	Zn	240
Fe	453	AgI	134
Ge	374	BN	1900
Hg	90	CuCl	207
K	99	GaAs	344
Li	430	ZnS	530

APPENDIX I

THERMODYNAMIC DATA FOR CHEMICAL REACTIONS

Values for the standard Gibbs free energy change for a number of reactions are tabulated below. The numbers represent empirical fits to the equation

$$\Delta G^0 = A + BT \ln T + CT, \tag{I.1}$$

and give ΔG^0 values in joules per mole. The parameter A represents the Gibbs free energy change for the reaction at 0 K.

THERMODYNAMIC DATA FOR CHEMICAL REACTIONS

Reaction	A	B	C
$2Al(s) + \frac{3}{2}O_2(g) = Al_2O_3(s)$	−1,676,990	−7.23	366.7
$2Al(l) + \frac{3}{2}O_2(g) = Al_2O_3(s)$	−1,697,700	−15.69	385.9
$2Be(s) + O_2(g) = 2BeO(s)$	−1,200,980	−9.64	234.8
$C(graphite) = C(diamond)$	−1,298		4.73
$C(s) + H_2(g) = CH_4(g)$	−69,120	22.26	−63.35
$C(s) + \frac{1}{2}O_2(g) = CO(g)$	−111,770		−87.7
$C(s) + O_2(g) = CO_2(g)$	−394,130		−0.84
$2CO(g) + O_2(g) = 2CO_2(g)$	−565,120		173.7
$2Co(s) + C(s) = Co_2C(s)$	16,535		−8.71
$2Co(s) + \frac{1}{2}O_2(g) = CoO(s)$	−476,560		155.2
$4Cu(s) + O_2(g) = 2Cu_2O(s)$	−339,070	−14.28	247.0
$4Cu(l) + O_2(g) = 2Cu_2O(s)$	−390,980	−14.28	285.5
$Fe(s) + \frac{1}{2}O_2(g) = FeO$	−259,500		62.58
$Fe(s) + Cl_2(g) = FeCl_2(s)$	−346,480	−12.68	212.8
$Fe(s) + Cl_2(g) = FeCl_2(l)$	−286,530		63.7
$Fe(s) + Cl_2(g) = FeCl_2(g)$	−105,700	41.8	−357.3
$Ge(s) + \frac{1}{2}O_2(g) = GeO(g)$	−228,560	−19.24	259.5
$H_2(g) + \frac{1}{2}O_2(g) = H_2O(g)$	−239,650	8.14	−9.25
$Mg(s) + \frac{1}{2}O_2(g) = MgO(s)$	−603,960	−5.36	142.1
$Mg(l) + \frac{1}{2}O_2(g) = MgO(s)$	−608,140	−0.44	112.8
$MgO(s) + CO_2(g) = MgCO_3(s)$	−117,630		170
$Mo(s) + O_2(g) = MoO_2$	−232,560		5.86
$Ni(s) + \frac{1}{2}O_2(g) = NiO(s)$	−244,560		98.53
$Si(s) + SiO_2(s) = SiO(g)$	669,930	25.07	−508.6
$Si(s) + O_2(g) = SiO(s)$	−881,150	−5.45	218.5

APPENDIX J

WORK FUNCTIONS AND FERMI ENERGIES

Substance	Work Function (eV)	Fermi energy (eV)
Ag	4.63	5.48
Al	4.28	11.63
Au	5.38	5.51
Cd	4.22	7.46
Cu	4.65	7.04
Ga	4.20	10.35
In	4.12	8.60
K	2.30	2.12
Mg	3.66	7.13
Na	2.75	3.23
Pb	4.25	9.37
Zn	4.33	9.39

Work function values are either (a) experimental values for polycrystalline materials or (b) averages over measurements on low-index planes of single crystals.

INDEX

Absolute temperature:
　definition, 13–14
　and β, 86
Absolute zero of temperature, 15
Activity, 219, 224, 311
　definition, 211–212
　in a solution, 227
Activity coefficient, 232
　at infinite dilution, 229
Adiabatic:
　process, 7
　wall, 9
Adsorption:
　gas, 255
　isotherm, 257
　sites, 250
Allowed wavelengths, 177
Arsenic (As_4), 198
Atomic:
　mobility, 87
　weight, 288
Avogadro's number, 214

Binding:
　energy, 166, 237
　sites, 235
Blackbody radiation, 331–339
Boiling points, tabulation, 344–345
Boltzmann's constant, 143
Boltzmann statistics, 112, 135, 136
Bond energy, 149

Born-Oppenheimer approximation, 149
Bose-Einstein statistics, 107, 112, 331
Bragg-Williams model, 233, 257, 291
　of nonideal lattice gas, 252–255
　of solid solution, 267–272

Canonical ensemble, 66–70, 79
　and β, 84–87
　definition, 59
　partition function, 70, 109
Carbon:
　gasification of, 310
　isotopes, 307
Carnot cycle, 13–14
Cell model of liquid, 187, 189, 277
Center-of-mass coordinate system, 151
Characteristic function, 23
Chemical equilibrium, see Equilibrium, chemical
Chemical potential, definition, 22
Chemical reactions, see Reactions, chemical
Chemical system, 20
Clapeyron equation, 130–132, 197, 201
Closed system:
　chemical, 86
　isobaric, 26
　isothermal, 25, 61, 78–79
Coefficient of volume thermal expansion, 31
Coexistence:
　curves, 128
　lines, 130–131

Collision diameter, 193
Communal entropy, see Entropy, communal
Component, 20
Compressibility, isothermal, see Isothermal compressibility
Concentrated solution, see Solution(s), concentrated
Configurational entropy, see Entropy, configurational
Constant(s):
 fundamental, 341
 k (Boltzmann's constant), 75
 β, 69
 γ, 71
Constituent, definition, 20
Continuous function, 32
Coordination number, 165, 203, 250-251
Copper, 327-329
Criteria for equilibrium, see Equilibrium, criteria for
Critical:
 point, 129, 198
 pressure, 123
 temperature, 127
  for lattice gas, 255
Cross-differentiation identity, 32
Crystalline solid, 110, 163
Crystal surface, 240

Debroglie:
 equation, 103, 105
 wavelength, 103, 104
Debye:
 function, 183
  tabulation of, 354-355
 model, 175-178, 310, 320, 332, 336
 temperature, 183
  of diamond, 184
  of lead, 184
  tabulation of, 355
 treatment, 102
Degeneracy, 55, 61, 144
 of energy levels, 106
 of rotational energy levels, 155
Degrees of freedom, 129, 143
 in Debye solid, 177
 definition, 40
 in polyatomic molecules, 158
Density:
 of liquid, 203
 of solid, 203
Density of states:
 in Debye solid, 178
 in electron gas, 328

Diathermic wall, 10
Diatomic:
 gas, 149-152
 molecule(s), 113, 149-157
  electronic contribution to energy, 152
  homonuclear, 156
  translational modes, 152
Dilute solutions, see Solution(s), dilute
Dissociation energy, 151
Distinguishability, 110
Distinguishable:
 atoms, 163
 particles, 109, 110
Distributive equilibrium, see Equilibrium, distributive
Dulong and Petit, law of, 173

Eigenvalues, 105
Einstein:
 crystal, 235, 263
  with vacancies, 236-240
 model, 163-166, 198, 310
 temperature, 168
 treatment, 102
Elastic:
 continuum, 175
 modulus, 184
Electron(s), 106
 free, 317
 gas, 103, 317-329
 spin states, 319
Electronic:
 excitation, see Excitation, electronic
 excited state, see Excited state, electronic
 ground state, see Ground state, electronic
Electron volt, 144
Energy:
 of adsorption, 243
 of fusion, 287
 level(s), 55, 61, 94
 of mixing, 269, 273
 states, 59, 94
 storage, modes of, 108
 of sublimation, 203
  Einstein solid, 172
Ensemble(s), 56-63
 average energy, 61
 average values, 58, 83
 canonical, see Canonical ensemble
 choice of, 97
 of combined systems, 68
 grand canonical, see Grand canonical ensemble
 method, 56

microcanonical, see Microcanonical
  ensemble
 other, 342-343
Enthalpy:
 definition, 22
 of fusion, 124
 of mixing, 232
 of vaporization, 125, 132
Entropy:
 change, calculation of, 13
 communal, 190, 193, 280
 configurational, 239
 definition, 13
 and the function S', 78
 of fusion, 4, 123, 190, 287
  tabulation, 344-345
 ideal gas, 143
 of mixing, 232, 266, 269, 274
 principle, 19-20, 23-24, 40
 of vaporization, 124
  tabulation, 344-345
 vibrational, 239
Equation of state, 142, 160
 definition, 6
Equilibrium:
 binding sites, 242
 chemical, 6, 47, 303-313
  criterion for, 44-46
 constant, 303-304, 311
  in terms of activities, 304
  in terms of pressures, 306
 criteria for, 5, 23, 75-81, 123
 distributive, 47, 71, 198
  adlayer-gas phase, 246
  criterion, 43-44
  two-component system, 283
 hydrostatic, 46
  criterion, 42-43
 interatomic distance, 150, 164
 isotopic, 307-309
 liquid-vapor, 198-200
 mechanical, 6
 metastable, 129
 partial presure, 127
 phase:
  one-component system, 197
  multicomponent system, 283-300
 solid-liquid:
  one-component system, 200-201
  two-component system, 284-290, 293-300
 solid-vapor, 197-198
 thermal, 6, 46
  and β, 69
  criterion, 42

vacancy concentration, 239-240
vapor presure, 4, 132, 197, 212
 empirical expression, 199, 346-347
 over liquid, 199
 over two-component solution, 290-291
 tabulation, 346-347
Eutectic:
 point, 298
 system, 299
Excess functions, 232-233
Excitation, electronic, 139, 144-145
Excited state:
 electronic, 153
 molecular, 153
Exclusion principle, 107
Extensive coordinate, 5

Face-centered cubic structure, 203, 204, 288
Fermi-Dirac statistics, 107, 112, 317-321
Fermi energy, 326, 328
 tabulation, 358
First law of thermodynamics, 12
First-order phase change, see Phase change,
  first-order
Fluctuations:
 magnitude of, 94-97
 in pressure, 57
Fluorine, 145
Force constant, see Vibrational motion, force
  constant
Free electrons, see Electrons, free
Free energy:
 and the function F', 78-79
 of mixing, 266, 269, 273, 274
 surfaces, 119-127
Free volume, 190
Fundamental constants, see Constants,
  fundamental

Gamma function, 141
Gas:
 adsorption, see Adsorption, gas
 constant, 3
Gas-phase reactions, see Reactions,
  gas-phase
Gaussian distribution, 95
Gibbs, J. Willard, 56
Gibbs-Duhem equation, 40-41, 47-48, 130,
  214-215
Gibbs free energy, 119
 definition, 22
 one-component system, 110-127
 phase rule, see Phase rule, Gibbs
 standard state, 120

Grand canonical ensemble, 62, 70–72, 158–160
   definition, 59–60
   and γ, 90–93
   partition function, 158–160, 320
      definition, 72
Ground state:
   electronic, 145, 153
   energy, 145
   vibrational, 168, 172

Half integral spins, see Spins, half integral
Halogens, 145
Harmonic oscillator model of liquid, 192
Heat:
   capacity, 318
      at constant pressure, 31
      at constant volume, 31
      electron gas, 329
   flow, 7, 12
   of fusion, tabulation, 344–345
   of vaporization, tabulation, 344–345
   reservoir, 67
Heisenberg uncertainty principle, 103
Helmholz free energy:
   definition, 22
   of ideal gas, 142
Henry's law, 229, 231
Heterophase reactions, see Reactions, heterophase
Hill, Terrell, 257
Homonuclear diatomic molecule, see Molecule, homonuclear diatomic
Hydrostatic equilibrium, see Equilibrium, hydrostatic
Hyperbolic functions, tabulation, 352–353

Ideal:
   entropy of mixing, see Entropy of mixing
   gas, 102, 132, 198
      monatomic, 103, 135–146, 212
      equation of state, 6
      mixtures, 224–226
   lattice gas, see Lattice gas, ideal
   solutions, see Solutions, ideal
Identical particles, 110
Independent particle systems, 108
Indistinguishable particles, 110, 111
Inexact differential, 11
Infinitesimal process, 7
Integral spin, see Spin, integral
Intensive coordinate, 5
Interacting particles, 249
Interaction:
   energy, 250–251, 254

Bragg-Williams model, 253
   potential, 150
Interatomic:
   distance, 277
   potential, 278
   spacing, 242
Internal energy:
   definition, 11
   ideal gas, 143
Interparticle interaction, 106–108
Ion core, 106
Irreversible process, 8, 19
Isolated system, 62
   criterion for equilibrium, 24–25, 75–78
Isothermal compressibility, 31, 201
Isotopic equilibrium, see Equilibrium, isotopic

Jellium model, 317–318

Langmuir:
   adsorption isotherm, 245–247, 254
   model of adsorption, 240–247
Lattice:
   gas, 235
      ideal, 235
      nonideal, 249–258
   model of solutions, see Solution(s), lattice model
   statistics, 235–247
   three-dimensional, 163
Lennard-Jones, J. E., 349
   potential, 150, 202, 349
   for liquid, 188
Linear triatomic molecule, see Molecule, linear triatomic
Liquid:
   models of, 187–196
   solutions, see Solutions, liquid
Liquidus temperature, 296
Liquid–vapor equilibrium, see Equilibrium, liquid–vapor
Longitudinal vibrational modes, 177

Macroscopic approach, 3–5
Macrostate, 6, 54
Maxwell's relations, 34–35
Mechanical:
   equilibrium, see Equilibrium, mechanical
   variables, 56, 75, 83
Melting point(s), 200, 284, 287–288, 329
   maximum, 296
   minimum, 296
   tabulation, 344–345
Metals, oxidation, 312

Metastable phase, *see* Phase, metastable
Microcanonical ensemble, 62, 66
　definition, 59
Microscopic approach, 3-5
Microstate, 54
　definition, 6
Model(s), 101-102
　ideal gas, 135
　potentials, *see* Potentials, model
　of a system, 58
Molar:
　entropy, 129
　properties, 212-213
　　definition, 129
Molecular excited state, *see* Excited state, molecular
Molecule(s):
　homonuclear diatomic, 304-307
　linear triatomic, 309
　nonlinear triatomic, 309
　polyatomic, 158, 309-310
Mole fraction, 211-212
Moment(s) of inertia, 155, 308
　polyatomic molecules, 158
Morse, P. M., 348
　potential, 348
Multicomponent system, 29, 211

Nearest neighbor:
　interactions, 262
　model, 165, 202
Nernst-Simon statement (of third law), 15
Nonideal gas, 102, 187
Nonlinear triatomic molecule, *see* Molecule, nonlinear triatomic
Nonmechanical variables, 56, 83
Number density (of atoms), 184

Open isothermal system, 26-27, 79-81
Ordering, 275
Other ensembles, *see* Ensembles, other
Oxidation of a metal, *see* Metals, oxidation
Oxygen isotopes, 307

Pair potential, 150, 262, 288, 292
　summation, for liquid, 188
Pairwise interactions, 164, 202
　in Langmuir model, 240-242
Partial derivatives, 29-30
Partial molar:
　Gibbs free energy, 213-214
　properties, 219, 233
　quantities, 213-219
Permittivty (of a vacuum), 106

Phase:
　change, 255
　closed, 20
　definition, 20
　diagram(s):
　　copper-nickel, 289-90
　　germanium-silicon, 290
　　one-component system, 201-206
　　two-component systems, 283, 289-290
　equilibrium, *see* Equilibrium, phase
　metastable, 189
　open, 20
　rule, Gibbs, 47-48, 129
　separation, 291-293, 296, 298
Phonon gas, 336
Phosphorous (P_4), 198
Photon(s), 331
　gas, 331-339
Planck's:
　constant, 103
　formula, 338-339
Polyatomic molecules, *see* Molecules, polyatomic
Postulates, 58
　application of, 65
　statement, 60
Potential(s):
　energy curve, 150
　field, 105
　model, 348-349
　well, 163
　　depth, 165, 172, 188-189
　　for liquid, 187
Pressure, 56
　ideal gas, 142
　instantaneous value, 56
　long time average, 56, 57
　two-dimensional, 243-245
Probability:
　compound, 67-68
　of occupancy (of state), 321

Quantum:
　harmonic oscillator, 154
　mechanics, 53, 102-108
　numbers, 136
　number space:
　　Debye solid, 176-177
　　electron gas, 319, 324
　　ideal gas, 136, 140
　　photon gas, 332
　rotator, 155
　state, 55, 60, 61. *See also* Energy state
　　probability of occurrence, 67

Quasichemical model, 233
  of lattice gas, 256–258
  of solid solution, 272–277
Quasi-static process, 7

Radial distribution function, 188–189, 203
Radiant heat transfer, 337
Random:
  solution, see Solution(s), random
  structure (of liquid), 187
Raoult's law, 228, 231
Rayleigh-Jeans formula, 339
Rayleigh, Lord, 339
Reaction(s):
  chemical (thermodynamic data), 356–357
  gas phase, 309–310
  heterophase, 310–313
  spontaneous, 46
  water-gas, 309
Reduced:
  mass, 154, 308
  temperature plot:
    Debye solid, 186
    Einstein solid, 170–171
Reference pressure, 212
Regular solution, see Solution(s), regular
Relative coverage, 244–247
Reservoir, 58
Reversible process, 7, 19
Rotational:
  energies, 105
  motion, 104–105
  temperature, 155

Sackur-Tetrode equation, 144
Schrödinger equation, 105, 107, 151
Second law of thermodynamics, 12–13
  statement, 13
Solid–liquid equilibrium, see Equilibrium, solid liquid
Solid–vapor equilibrium, see Equilibrium, solid–vapor
Solid solutions, see Solution(s), solid
Solidus temperature, 296
Solution(s):
  classical treatment, 226–233
  concentrated, 232
  dilute, 229–232
  ideal, 228–229
  lattice model, 261–263
  liquid, 277–281
    ideal, 279–280
  random, 271
  regular, 232, 252, 267–272, 280–281
  solid, 263–277
    ideal, 264–267
    statistical thermodynamic treatment, 261–281
    two-component, 261–281
Spectral energy distribution, 337–339
Spectroscopic:
  data (diatomic and polyatomic molecules), 350–351
  measurements, 3
Spin:
  half integral, 107
  integral, 107
Standard:
  deviation, 95
  free energy change (in reaction), 304
  free energy of reaction, 311
  state:
    presure, 305
    Gibbs free energy, 120
Standing waves:
  in photon gas, 331, 332, 336
  in Debye solid, 175, 176
State function(s), 22
  evaluation in terms of Q, 87–89
  evaluation in terms of Ξ, 93–94
  equations (integration of), 35
Statistical mechanics, see Statistical thermodynamics
Statistical thermodynamics (basis of), 53–55
Stirling's approximation, 141, 237, 243, 253
Stoichiometric coefficients, 303
Stoichiometry, 45
Surface area, 23
Symmetry number, 156
System energies (allowed values), 102

Thermal equilibrium, see Equilibrium, thermal
Thermal expansion coefficient, 201, 262
Thermally conducting wall, 9, 10
Thermodynamic:
  coordinate, 5
  state, 5
  functions:
    Debye solid, 181–182
    diatomic molecule, 156–157
    Einstein solid, 169–170
    electron gas, 322–326
    photon gas, 335–337
  system, 5
Third law of thermodynamics, 15
Time-average values, 83
Total molar properties, 219
Transcendental equations, 295

INDEX   365

Translational:
　energy states, 136
　motion, 103, 138
Transverse vibrational modes, 177
Triple point, 128, 198, 200, 204
Two-component:
　phase diagram, *see* Phase diagram(s),
　　two-component
　system(s):
　　ideal, 284–291
　　nonideal, 291–300
Two-dimensional pressure, *see* Pressure,
　two-dimensional

Unattainability statement (of third law), 15
Unit conversions, 341
Unoccupied sites, 251

Vacancy, 236–240
　concentration, 238, 240
Valence electrons, 318
van der Waals:
　equation of state, 6
　gas, 107
Vapor pressure, *see* Equilibrium vapor
　pressure
Velocity:
　of light, 332
　of propagation (of wave), 176
Vibrational:
　energy, 105
　entropy, *see* Entropy, vibrational
　excitation (diatomic molecule), 155
　frequency(ies):
　　diatomic molecule, 308

　　Einstein solid, 167
　　polyatomic molecule, 158
　modes (Debye solid), 177
　motion:
　　of adsorbed atom, 242
　　of atoms, 105
　　force constant, 166
　　in nonideal lattice gas, 252
　temperature, 154
Volume change:
　on fusion, 127
　　tabulation, 344–345
　on mixing, 262
　　ideal solution, 229
　on vaporization, 127

Water-gas reaction, *see* Reaction(s), water-gas
Wien:
　displacement law, 339
　formula, 339
Wave function, 105
　antisymmetric, 107
　symmetric, 107
Well depth (maximum), 150
Work, 7
　definition, 10
　function, 326
　　tabulation, 358
　path dependence, 10

Zero'th law of thermodynamics, 9–10, 69
　statement, 10
Zero of energy, 106, 150, 153, 188, 241
　definition, 145
Zeta function, 333